T0234535

NAUTICAL MATHEMATICS

by

S. A. WALLING
Senior Master R.N. (Ret.)

and

J. C. HILL, B.A. (CANTAB.)
Education Department, Cambridge University Press

CAMBRIDGE
AT THE UNIVERSITY PRESS
1943

CAMBRIDGE
UNIVERSITY PRESS

University Printing House, Cambridge CB2 8BS, United Kingdom

Published in the United States of America by Cambridge University Press, New York

Cambridge University Press is part of the University of Cambridge.

It furthers the University's mission by disseminating knowledge in the pursuit of education, learning and research at the highest international levels of excellence.

www.cambridge.org
Information on this title: www.cambridge.org/9781107642522

First published 1943
First paperback edition 2014

A catalogue record for this publication is available from the British Library

ISBN 978-1-107-64252-2 Paperback

CONTENTS

FOREWORD

by

REAR-ADMIRAL J. G. P. VIVIAN, C.B.

Admiral Commanding Reserves

A sound knowledge of mathematics is essential to the seaman; without that knowledge he is neither capable of being a competent "ship's husband" nor of navigating his ship across oceans and seas or in coastal waters.

The dry bones of mathematics have here been made alive and then clothed with a seaman's rig.

To you who have sea sense and the urge to go down to the sea in ships it is an exciting book, for it shows you how to solve many of those innumerable problems with which you will be confronted by day and by night, at sea and in harbour, in every type of ship.

Whether you be a deck hand or the captain of the ship, you will find that the knowledge which this book can impart is the basis on which the material efficiency of your ship and, that which is of equal importance, the safety, comfort and well-being of your shipmates, will largely depend.

These two things go together, the efficient ship and the happy ship's company, and both are dependent on the knowledge and skill of the leaders.

I heartily commend this book to you who aspire to be leaders in the Sea Services; if you use it with intelligence and with diligence you will be on the high road to achieve that aspiration.

PREFACE

The call of the sea is in the blood of many of the youth of this country and the enthusiastic enrolment in such voluntary organisations as the Sea Cadet Corps is excellent evidence of this.

It is the object of those to whom the instruction of these cadets is entrusted to encourage them, by careful training and example, to fit themselves for positions of responsibility and authority in whatever branch of the Sea Services they may ultimately serve.

A most necessary preliminary to the intelligent understanding of many of the activities and duties of a sailor's life is a sound knowledge of the elementary principles of mathematics and their applications.

Much of this will have been done while at school, but for the benefit of those who have become "rusty" by lapse of time and as a stimulant to those who found difficulty or distaste in the academic study of figures, this book has been carefully designed.

Only as little as is considered essential for revision and accuracy in calculation is of a mechanical nature. The rest, without exception, has been compiled to apply these fundamental rules to the many interesting and important problems of seamanship.

It is hoped that this "flavour of the sea" may encourage the cadet and offer him a more acceptable road to proficiency.

The usefulness of this book should be, by no means, confined to members of the Sea Cadet Corps. Those already in the Royal Navy or the Merchant Service desiring to fit themselves for promotion or to specialise in some particular branch may find much here to help them.

The authors wish to express their very sincere thanks to Commander W. T. Marsh, R.N.V.R., with whose keen interest and assistance this book was originally planned; to Instructor Rear-Admiral A. E. Hall, C.B., C.B.E. and the Controller of H.M. Stationery Office for permission to include the Naval Educational Tests, and to Rear-Admiral J. G. P. Vivian, C.B. (Admiral Commanding Reserves), who has been kind enough to examine the manuscript and write the Foreword.

S. A. W.
J. C. H.

November 1942

TABLES

Length.

12 inches (in.)	= 1 foot (ft.)
3 feet	= 1 yard (yd.)
6 feet	= 1 fathom
100 fathoms	= 1 cable = 8 shackles
10 cables	= 1 nautical mile of 2000 yd.

This measure (2000 yd.) of a nautical mile is only approximate, although frequently used at sea. A more accurate measure of a nautical mile is the length of an arc of the equator which subtends an angle of 1 minute (1′) at the centre of the earth. This length is very nearly 6080 feet, which length is known as the *Mean Nautical Mile.*

1 statute mile = 5280 feet.

Time.

60 seconds (sec.)	= 1 minute (min.)
60 minutes	= 1 hour (hr.)
24 hours	= 1 day
7 days	= 1 week
4 weeks	= 1 lunar month
365 days	= 1 year
366 days	= 1 leap year

Measurement of Angles.

60 seconds (″)	= 1 minute (′)
60 minutes	= 1 degree (°)
60 degrees	= 1 sextant
90 degrees	= 1 quadrant (or right angle)
360 degrees	= 1 full circle

Weight.

16 ounces (oz.)	= 1 pound (lb.)
14 pounds	= 1 stone (st.)
28 pounds	= 1 quarter (qr.)
112 pounds	= 1 hundredweight (cwt.)
20 hundredweights	= 1 ton

THE FOUR RULES

General Revision of Addition, Subtraction, Multiplication and Division of Numbers.

Every calculation that has to be made, to obtain the answer to a problem, involves one or more of the above four rules.

It is essential, therefore, that accuracy in the use of numbers must be the first and most important consideration.

Always check an answer, and wherever possible do this by an alternative method.

For example, when working an addition sum add the numbers a second time in a different order, so that any mistake made during the first calculation is less likely to be made in the second.

When multiplying numbers, change the order of multiplication the second time for the same reason.

Addition and Subtraction of Numbers.

Exercise I.

1. 456 + 789 + 120.
2. 246 + 804 + 468.
3. 779 + 416 + 321.
4. 1395 + 476 + 2764.
5. 6959 + 2175 + 5432.
6. 30,142 + 2731 + 108 + 94,615.
7. 214 + 3886 + 2994 + 14,095.
8. 5608 + 300 + 4215 + 71,230.
9. 399 + 1387 + 3796 + 14,895.
10. 4594 + 4378 + 374 + 2806.
11. 273 − 241.
12. 496 − 454.
13. 524 − 217.
14. 694 − 589.
15. 3762 − 1759.
16. 14,360 − 9079.
17. 10,051 − 9896.
18. 22,812 − 19,394.
19. 24,005 − 2998.
20. 10,096 − 9997.

Multiplication and Division of Numbers.

Exercise II.

1. 896 × 89.
2. 789 × 96.
3. 587 × 198.
4. 1756 × 596.
5. 3417 × 8370.
6. 514 × 213 × 17.
7. 91 × 626 × 727.
8. 734 × 295 × 317.
9. 731 × 195 × 263.
10. 478 × 677 × 954.
11. 2544 ÷ 47.
12. 2136 ÷ 89.
13. 12,673 ÷ 29.
14. 95,157 ÷ 97.
15. 124,944 ÷ 456.
16. 1,067,520 ÷ 278.
17. 604,298 ÷ 838.
18. 115,189 ÷ 4266.
19. 395,638 ÷ 4986.
20. 274,321 ÷ 2878.

Addition, Subtraction, Multiplication and Division of Quantities.

A quantity describes a number of some definite thing or unit.

For instance, 17 tons, 12 seconds, 25 yards are quantities.

Revision of Length.

EXERCISE III.

1. Find the sum of 2 yd. 1 ft. 9 in., 5 ft. 7 in., 6 yd. 0 ft. 11 in.
2. Find the sum of 3 fathoms 3 ft. 10 in., 10 ft. 9 in.
3. Express 5 ft. + 12 ft. 5 in. − 8 ft. 9 in. in feet and inches.
4. Express 3 fathoms + 5 ft. 8 in. in inches.
5. Express 1209 in. in fathoms, feet and inches.
6. Express 3 shackles 12 yd. 1 ft. 6 in. in yards, feet and inches.
7. Express 2 cables 4 shackles in feet.
8. Express 1 cable 5 shackles in yards.
9. 4 ft. 6 in. × 8.
10. 3 ft. 2 in. × 7.
11. 1 yd. 2 ft. 3 in. × 6.
12. 2 fathoms 2 ft. 4 in. × 4.
13. 3 shackles 3 ft. 3 in. × 11.
14. 5 fathoms 4 ft. 6 in. × 19.
15. Divide 8 ft. 2 in. into 7 equal parts.
16. 7 yd. 1 ft. ÷ 8.
17. 4 fathoms 0 ft. 6 in. ÷ 7.
18. 2 shackles 5 ft. ÷ 31.
19. 264 fathoms 1 ft. 6 in. ÷ 21.
20. 11 yd. 0 ft. 7 in. ÷ 13.

Revision of Time.

EXERCISE IV.

1. 2 hr. 20 min. 25 sec. + 5 hr. 15 min. 13 sec. + 3 hr. 41 min. 46 sec.
2. 3 hr. 11 min. 8 sec. − 1 hr. 38 min. 25 sec.
3. 4 hr. 19 min. 17 sec. × 8.
4. 16 hr. 8 min. 12 sec. ÷ 12.
5. Express 2 hr. 46 min. 20 sec. in seconds.
6. Express 10,000 sec. in hours, minutes and seconds.
7. How many hours are there between 6.20 a.m. on 10 April to 4.20 a.m. on 12 April?
8. How many hours are there in a leap year?
9. How many days are there from 1 July to 31 December inclusive, in any year?
10. How many days are there from 1 January to 31 March inclusive, in a leap year?

Revision of Weight.

Exercise V.

1. 2 tons 13 cwt. 3 qr. + 7 tons 10 cwt. 2 qr. + 19 tons 17 cwt.
2. 25 tons 8 cwt. 2 qr. − 17 tons 15 cwt. 3 qr.
3. Express 3 tons 17 cwt. 2 qr. in stones.
4. Express 1589 cwt. in tons and hundredweights.
5. 5 tons 17 cwt. 3 qr. × 16.
6. 25 tons 3 cwt. 3 qr. ÷ 13.
7. 4 tons 17 cwt. − 2 tons 19 cwt. + 6 tons 15 cwt.
8. 11 st. 13 lb. + 12 st. 10 lb. + 10 st. 12 lb.
9. Express 5 cwt. 2 st. in pounds.
10. Express 2128 lb. in hundredweights.

Revision of Angles.

Exercise VI.

1. $24° 16' 20'' + 32° 58' 43'' + 19° 32' 27''$.
2. $38° 12' 21'' - 19° 36' 46''$.
3. $5° 17' 32'' \times 9$.
4. $110° 42' 10'' \div 11$.
5. Express $3° 28' 41''$ in seconds.
6. Express 19,680″ in degrees, minutes and seconds.
7. By how much is $79° 59' 59''$ short of one right angle?
8. By how much is the sum of $23° 10'$ and $110° 20'$ short of two right angles?
9. Express $7 \times 17'$ in degrees and minutes.
10. By how much is $235° 22' 5'' \div 5$ short of a sextant?

Problems on the Four Rules.

Unless otherwise stated, assume 1 nautical mile to be equal to 6080 feet. The abbreviation for nautical mile is N.M. A speed of 1 N.M. per hour is called a *knot*.

Exercise VII.

1. A ship, on five successive days, made the following runs in nautical miles: 243, 259, 282, 279 and 286. What was the total distance steamed in the five days?

2. A ship steamed 1004 N.M. in 4 days. During the first 2 days she covered 236 and 254 N.M. respectively. What was the combined run during the last two days?

3. A vessel left Plymouth bound for Alexandria, a distance of 2872 N.M. The distance from Plymouth to Lisbon was recorded as 770 N.M.; from Lisbon to Gibraltar as 302 N.M.; and from Gibraltar to Malta as 980 N.M. What is the distance from Malta to Alexandria?

4. Two ships leave Southampton for Melbourne, Australia. The first takes the Cape route and records the following distances:

Southampton to Cape Town	5947 N.M.
Cape Town to Melbourne	5814 N.M.

The second ship takes the Red Sea route, with the following log records:

Southampton to Gibraltar	1172 N.M.
Gibraltar to Suez	1920 N.M.
Suez to Aden	1310 N.M.
Aden to Colombo	2130 N.M.
Colombo to Freemantle	3120 N.M.
Freemantle to Melbourne	1553 N.M.

Which ship makes the shorter passage and by how many nautical miles?

5. Before the building of the Panama Canal all ships from Eastern American ports bound for Western American ports had to sail via South American waters and Cape Horn. What is the saving in a journey from New York to San Francisco if the distance from New York to Panama is 1985 N.M. and from Panama to San Francisco is 3260 N.M., while the distance from New York to Montevideo is 5723 N.M. and from Montevideo to San Francisco is 7536 N.M.?

6. What distance is saved between Southampton and Colombo by taking the Suez Canal route?

The distance from Southampton to Suez is 3092 N.M. and from Suez to Colombo is 3440 N.M. The distance from Southampton to Cape Town is 5947 N.M. and from Cape Town to Colombo is 4438 N.M.

7. A vessel steams a steady 18 knots for 36 hr. How far has she travelled in this time?

8. If a ship records a steady noon to noon run of 272 N.M. for 17 days, how far has she travelled?

9. From the following table of entries in a ship's speed log, find the total distance run between noon on 10 May 1942 and 8.0 a.m. on 13 May 1942:

Date	Time	Speed	N.M.
May 10	12 00	13 knots	—
10	17 00	16	65
11	8 00	15	
12	12 00	12	
13	8 00		
		Total distance run	

10. Which is the greater distance, 7 statute miles or 6 nautical miles? What is the difference in feet?

11. What is the difference, in feet, between a distance of 66 nautical miles and 76 statute miles?

12. The Russian unit of length is the verst, and 1 verst = 3500 English feet. Which is the greater distance, and by how many feet, 32 versts or 22 statute miles?

13. Which is the greater distance, and by how many feet, 35 versts or 21 nautical miles?

14. A straight line of mooring buoys is laid at regular intervals of 3 cables. What is the distance in yards from the first buoy to the fifth buoy?

15. If a vessel steamed 5061 N.M. in 21 days at a constant speed throughout, what was the distance covered each day?

16. In 43 hr., steaming at a steady speed, a ship covered 1075 N.M. What was her speed in knots?

17. The following table shows how to determine the approximate tonnage of any number of bags of grain of different weights:

Lb. of grain per bag	No. of bags (N)	Tonnage
138	N	$N \div 15$
163	N	$N \times 4 \div 55$
168	N	$N \times 3 \div 40$
196	N	$N \times 7 \div 80$
207	N	$N \times 6 \div 65$
216	N	$N \times 3 \div 31$

A ship loads 1100 bags of grain each weighing 163 lb. What is the weight of her cargo?

18. A vessel discharges 4875 bags of grain each weighing 138 lb. By what weight has her cargo been reduced?

19. A ship's cargo consists of 960 bags of grain each weighing 168 lb., and 1625 bags each weighing 207 lb. What tonnage does this cargo represent?

20. A vessel discharges 1240 bags of wheat each weighing 216 lb. and loads 1440 bags of rice each weighing 196 lb. What is the alteration in the weight of her cargo?

21. If one ton of graded Newcastle coal occupies a space of 43 cu. ft., how many cubic feet of bunker space must be available in a ship taking on 350 tons of this coal as fuel?

22. A battleship carries 3 Bower anchors each weighing 125 cwt., 1 stream anchor weighing 42 cwt., and 3 kedge anchors, one of which weighs 15 cwt. and the other two 5 cwt. each. What, in tons and hundredweights, is the total weight of the anchors carried?

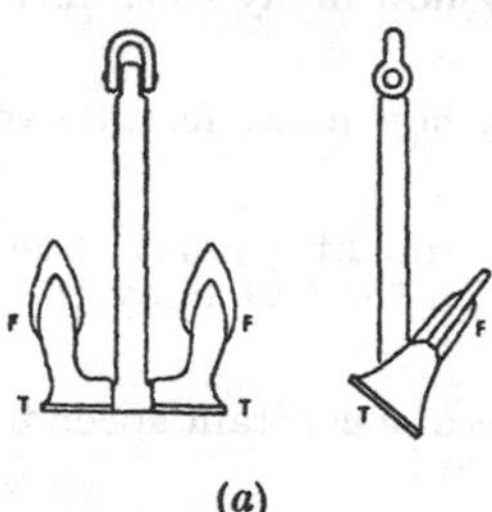

(*a*)
Stockless Anchor
(Bower Anchor)

The flukes *F* are hinged on the shank and are tripped by the tripping palms *T*.

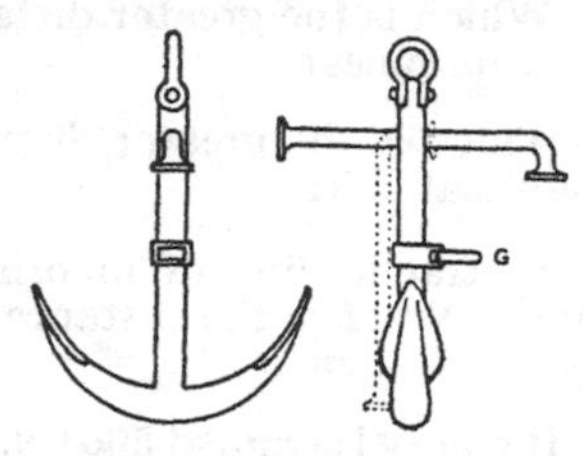

(*b*)
Admiralty Pattern
(Stream and Kedge Anchors)

The anchor is hoisted in board by means of the gravity band *G*.

23. A cargo ship is loading 2600 bags of wheat each weighing 207 lb., 400 bags of barley each weighing 196 lb., and 2040 bags of maize each weighing 168 lb. What weight in tons must be placed equally to port and starboard of the ship's fore and aft line to preserve balance? (Use the table in question 17.)

24. What is the total weight of 150 fathoms of anchor cable weighing 2 tons 10 cwt. per shackle?

FACTORS AND LEAST COMMON MULTIPLE

Factors.

If a number divides exactly into another number, the first number is said to be a *factor* of the second.

EXAMPLE 1. 3 is a factor of 21. So is 7 a factor of 21, and since $3 \times 7 = 21$, the factors of 21 are 3 and 7.

Many numbers have more than two factors, such as

$$24 = 2 \times 12 = 2 \times 2 \times 6 = 2 \times 2 \times 2 \times 3.$$

Any number that has no factors, other than itself and 1, is called a *prime number*, e.g. 2, 3, 5, 7, 11, etc.

It follows, therefore, that the factors of any number may be expressed as prime numbers.

EXAMPLE 2. $182 = 2 \times 91 = 2 \times 7 \times 13$, each of the final factors being a prime number.

These factors are called *prime factors.*

Least Common Multiple (L.C.M.).

A number is said to be a *multiple* of any of its factors.

Thus 180, 120 and 60 are all multiples of 20. They are also multiples of 12 and are consequently *common multiples* of 20 and 12.

The *smallest* number which is a common multiple of any factors is called the *least common multiple* (L.C.M.).

The L.C.M. of 20 and 12 is 60.

To determine the L.C.M. of two or more numbers it is necessary to find the factors of each of the numbers. The principle employed, to simplify this process and to avoid unnecessary work, is shown in the following examples.

EXAMPLE 1. Find the L.C.M. of 3, 5, 9, 14, 15, 21, 24.

First cross out any of the numbers which are factors of any of the others, thus:

~~3~~, ~~5~~, 9, 14, 15, 21, 24.

3, being a factor of 9, 15, 21 and 24, and 5, being a factor of 15, are crossed out, since any number which is divisible by 15 is necessarily divisible by 3 and 5.

Now divide the numbers by any prime factor common to two or more of the numbers, beginning with the lowest in order, 2, 3, 5, 7, 11, etc.

In this case divide by 2, and we have

2	~~3~~, ~~5~~, 9, 14, 15, 21, 24
	. . 9, 7, 15, 21, 12

Bring down any numbers, 9, 15 and 21 in this case, which are not divisible by the factor chosen.

Again cross out any number which is a factor of any of the others, thus:

2	~~3~~, ~~5~~, 9, 14, 15, 21, 24
	. . 9, ~~7~~, 15, 21, 12

Divide again by the next common prime factor (3 in this case) bringing down again any numbers which are not divisible by this factor:

2	~~3~~, ~~5~~, 9, 14, 15, 21, 24
3	. . 9, ~~7~~, 15, 21, 12
	. . 3, . 5, 7, 4

The numbers remaining have no common factor, so that the L.C.M. is the product of these numbers and the two factors by which the original numbers have been divided, i.e. the L.C.M. is

$$2 \times 3 \times 3 \times 5 \times 7 \times 4 = 2520.$$

Example 2. What is the shortest length of rope, in fathoms, which may be cut into lengths of 2, 3, 10, 12 or 18 fathoms with, in each case, no remainder?

In other words, what is the L.C.M. of 2, 3, 10, 12 and 18.

By the same method as in Example 1, we have

2	2, 3, 10, 12, 18
3	. . 5, 6, 9
	. . 5, 2, 3

The L.C.M. is, therefore, $2 \times 3 \times 5 \times 2 \times 3 = 180$. Therefore the shortest length of rope required to fulfil the above conditions is 180 fathoms.

Exercise VIII.

1. Find the prime factors of the following numbers:

(*a*) 16. (*b*) 45. (*c*) 85. (*d*) 168. (*e*) 251. (*f*) 9261.

2. Find the L.C.M. of:

(*a*) 18 and 27. (*b*) 48 and 60. (*c*) 6, 18, 42.
(*d*) 9, 54, 72. (*e*) 12, 16, 18, 27, 32. (*f*) 11, 66, 88, 99.

3. What is the shortest length of rope that can be cut exactly into lengths of either 10, 12, 15, 18 or 21 fathoms respectively?

4. What is the smallest number that can be an exact number of either dozens, scores or gross?

5. What is the smallest quantity of liquid, in gallons, that can be measured exactly with either a 2, 3, 5 or 12 gallon measure?

6. What is the smallest quantity of liquid, in gallons, that can be measured exactly with either a 4, 8, 12 or 15 pint measure?

7. The following are the weights of a cubic foot of different timbers: cedar 30 lb.; sycamore 25 lb.; elm 36 lb.; ash 45 lb.; spruce 32 lb. What is the smallest load in tons, cwt., lb. that could be an exact number of cubic feet of any of the above timbers?

8. The following are the weights of a cubic foot of different varieties of pine: yellow pine 32 lb.; red pine 36 lb.; white pine 27 lb.; Dantzig pine 40 lb. What is the smallest weight that would represent an exact number of cubic feet of any of these four timbers, and how many cubic feet of each would there be?

9. What is the shortest distance, in yards, that can be expressed as an exact number of statute miles or nautical miles (6080 ft.)?

10. What is the shortest length of chain, in feet, that could be made up of links all of which could be any one of the following inside lengths: 4 in., 6 in., 10 in., 1 ft., 1ft. 2 in., 1 ft. 3 in.?

FRACTIONS

A fraction is always a part of something. For example: three farthings are written as $\frac{3}{4}d.$ to denote that they are three-quarters of a penny.

In all fractions the number above the line is called the *numerator*, and that below the line is called the *denominator*.

All fractions should be expressed in their *lowest terms* and this is done by dividing the numerator and denominator by any common factors:

$$\frac{\overset{4}{\cancel{16}}}{\underset{5}{\cancel{20}}}=\frac{4}{5}.$$

All fractions in which the numerator is smaller than the denominator are termed *proper fractions.*

Fractions having a numerator greater than the denominator are called *improper fractions* and should never be left as such in an answer. They should be changed to a *mixed number* and reduced to their lowest terms. Thus

$$\frac{21}{9}=2\frac{3}{9}=2\frac{1}{3}.$$

It is frequently necessary when working with fractions to change mixed numbers into improper fractions, but they must not be left as such in the final answer.

Exercise IX.

1. Reduce the following fractions to their lowest terms:

(*a*) $\frac{6}{8}$. (*b*) $\frac{12}{16}$. (*c*) $\frac{18}{30}$. (*d*) $\frac{27}{81}$. (*e*) $\frac{28}{49}$.
(*f*) $\frac{45}{100}$. (*g*) $\frac{370}{555}$. (*h*) $\frac{216}{243}$. (*i*) $\frac{220}{924}$. (*j*) $\frac{1008}{1728}$.

2. Express the following mixed numbers as improper fractions:

(*a*) $1\frac{3}{4}$. (*b*) $3\frac{1}{2}$. (*c*) $8\frac{1}{7}$. (*d*) $6\frac{4}{5}$. (*e*) $4\frac{57}{100}$.
(*f*) $5\frac{3}{100}$. (*g*) $5\frac{7}{22}$. (*h*) $13\frac{2}{15}$. (*i*) $17\frac{7}{31}$. (*j*) $25\frac{2}{25}$.

3. Change these improper fractions into mixed numbers:

(*a*) $\frac{5}{4}$. (*b*) $\frac{8}{3}$. (*c*) $\frac{31}{2}$. (*d*) $\frac{17}{12}$. (*e*) $\frac{22}{7}$.
(*f*) $\frac{507}{100}$. (*g*) $\frac{630}{24}$. (*h*) $\frac{1143}{27}$. (*i*) $\frac{675}{72}$. (*j*) $\frac{402}{54}$.

4. Fill in the following blank spaces:

(*a*) $\frac{5}{6}=\frac{\quad}{24}$. (*b*) $\frac{2}{3}=\frac{\quad}{36}$. (*c*) $\frac{5}{9}=\frac{55}{\quad}$.
(*d*) $\frac{6}{17}=\frac{18}{\quad}$. (*e*) $\frac{3}{7}=\frac{\quad}{21}=\frac{\quad}{49}$. (*f*) $\frac{9}{10}=\frac{\quad}{20}=\frac{\quad}{100}$.
(*g*) $7=\frac{\quad}{3}=\frac{63}{\quad}$. (*h*) $13=\frac{\quad}{2}=\frac{\quad}{7}=\frac{117}{\quad}$.

Addition and Subtraction of Fractions.

When adding or subtracting fractions with different denominators, first obtain the L.C.M. of all the denominators.

The fractions can then all be expressed as fractions with the same denominator, by dividing each denominator into the L.C.M. and multiplying the answer, so obtained, by the numerator in each case.

EXAMPLE 1. Simplify $\frac{3}{8}+\frac{2}{3}-\frac{5}{6}$.

The L.C.M. of 8, 3 and 6 is 24. Thus

$$\frac{3}{8}+\frac{2}{3}-\frac{5}{6}=\frac{9}{24}+\frac{16}{24}-\frac{20}{24}$$

$\left(\text{this is usually written, for simplicity, in this form: } \frac{9+16-20}{24}\right)$

$$=\frac{25-20}{24}=\frac{5}{24}.$$

When adding or subtracting mixed numbers, do not make improper fractions. Merely find the total of the whole numbers and add this to the fractional answer, reduced to its lowest terms.

EXAMPLE 2. Simplify $4\frac{7}{10}-2\frac{11}{15}$.

$$4\frac{7}{10}-2\frac{11}{15}=2+\frac{7}{10}-\frac{11}{15}$$

(L.C.M. of denominators is 30)

$$=2+\frac{21-22}{30}.$$

We cannot subtract 22 from 21, so we borrow 1 or $\frac{30}{30}$ from the 2 leaving 1 as the whole number:

$$=1+\frac{30+21-22}{30}=1+\frac{51-22}{30}=1\frac{29}{30}.$$

EXERCISE X.

1. Find the value of:

(a) $\frac{2}{5}+\frac{3}{10}$. (b) $\frac{3}{14}+\frac{5}{21}$. (c) $\frac{4}{15}+\frac{3}{10}$.
(d) $\frac{5}{12}+\frac{7}{15}$. (e) $\frac{4}{11}+\frac{2}{33}+\frac{5}{22}$. (f) $5\frac{3}{8}+3\frac{5}{12}$.
(g) $1\frac{2}{5}+1\frac{1}{4}+\frac{7}{10}$. (h) $\frac{8}{9}+3\frac{5}{6}+1\frac{11}{12}$. (i) $1\frac{2}{15}+1\frac{19}{20}+1\frac{2}{3}$.
(j) $\frac{11}{24}+1\frac{7}{8}+1\frac{7}{10}$.

2. Find the value of:

(a) $\frac{1}{3}-\frac{1}{7}$. (b) $\frac{3}{4}-\frac{1}{6}$. (c) $\frac{7}{15}-\frac{3}{10}$. (d) $\frac{7}{12}-\frac{7}{18}$.
(e) $3\frac{5}{6}-2\frac{1}{3}$. (f) $3\frac{5}{8}-1\frac{3}{4}$. (g) $4\frac{3}{8}-2\frac{7}{12}$. (h) $7\frac{1}{2}-3\frac{3}{5}$.
(i) $6\frac{4}{7}-5\frac{3}{5}$. (j) $5\frac{3}{7}-4\frac{1}{2}$.

3. Simplify:

(a) $\frac{1}{4}+\frac{1}{2}-\frac{2}{5}$. (b) $1\frac{1}{10}-1\frac{1}{100}+1\frac{1}{1000}$. (c) $4\frac{1}{2}+2\frac{1}{6}-2\frac{1}{3}$.
(d) $4\frac{1}{8}-2\frac{1}{4}+4\frac{3}{4}$. (e) $7\frac{7}{12}-2\frac{17}{18}+3\frac{1}{6}$. (f) $3\frac{7}{15}-1\frac{13}{20}+\frac{17}{25}$.
(g) $6\frac{1}{2}-4\frac{1}{6}+3\frac{1}{12}-3\frac{3}{4}$. (h) $3\frac{5}{9}-2\frac{1}{3}+1\frac{5}{12}-1\frac{5}{6}$. (i) $6\frac{3}{22}-2\frac{17}{33}-1\frac{1}{2}$.
(j) $\frac{11}{14}+2\frac{1}{9}-\frac{3}{7}+5\frac{1}{6}$.

4. Take the smaller from the greater of:

(a) $\frac{5}{6}$ and $\frac{4}{7}$. (b) $\frac{77}{1000}$ and $\frac{2}{25}$. (c) $\frac{7}{8}$ and $\frac{7}{9}$.
(d) $7\frac{5}{6}$ and $7\frac{7}{8}$. (e) $\frac{8}{21}$ and $\frac{11}{28}$. (f) $\frac{15}{26}$ and $\frac{29}{39}$.

Problems involving Addition and Subtraction of Fractions.

EXERCISE XI.

1. Five rods are of the following diameters:

$\frac{2}{3}$ in.; $\frac{5}{6}$ in.; $\frac{3}{4}$ in.; $\frac{4}{5}$ in.; $\frac{7}{10}$ in.

Arrange them in order of magnitude, placing the largest first.

2. What would bolts of the following diameters measure if tested with a vernier micrometer reading thousandths of an inch?

(*a*) $\frac{3}{8}$ in. (*b*) $\frac{1}{4}$ in. (*c*) $\frac{7}{8}$ in. (*d*) $\frac{5}{16}$ in.

3. $\frac{2}{5}$ of a vessel's cargo is wheat, $\frac{1}{3}$ of the cargo is barley and the remainder oats. What fraction of the ship's cargo is oats?

4. A coil of rope is 122 fathoms long. What length, in feet and fractions of a foot, will be left after all the following issues have been made:

$4\frac{1}{2}$ ft., $5\frac{2}{3}$ ft., $16\frac{1}{2}$ ft., $100\frac{3}{4}$ ft., $72\frac{5}{12}$ ft., $32\frac{1}{4}$ ft.?

5. A fully laden cargo ship discharged $\frac{3}{8}$ of her cargo tonnage at one port and $\frac{1}{3}$ at another port. She then took on board $\frac{2}{5}$ of her total cargo tonnage. To what fraction of her tonnage capacity was she then loaded?

6. A vessel used $\frac{5}{16}$ of her full fuel load between Cape Town and Durban, and another $\frac{3}{5}$ between Durban and Zanzibar. What fraction of the ship's full fuel load remained?

7. A ship's speed was 12 knots for $\frac{5}{8}$ of its voyage, 13 knots for $\frac{1}{12}$ of the voyage and 14 knots for the remainder. What fraction of the voyage was made at 14 knots?

8. Of a ship's company $\frac{1}{4}$ are English, $\frac{1}{6}$ Scottish, $\frac{3}{16}$ Scandinavian, $\frac{1}{12}$ Dutch and the remainder Welsh. What fraction of the ship's company is Welsh?

9. A tide pole is used to indicate the depth of water at a place at different states of the tide. If $\frac{7}{8}$ of the pole is submerged at high water and $\frac{3}{5}$ of the pole is submerged at half tide, what fraction of the pole will be submerged at low water?

10. Of a certain alloy $\frac{4}{7}$ of its total weight is copper, $\frac{1}{3}$ is zinc and the remainder is tin. What fraction of this alloy is the weight of tin?

11. In a certain ship the port and starboard coal bunkers are each of the same capacity. The port bunker is completely full and the starboard bunker is $\frac{3}{5}$ full. If $\frac{2}{7}$ of the coal from the port bunker is transferred to the starboard bunker, which bunker will contain the more coal and by what fraction of the ship's total bunker capacity?

12. A vessel leaves port *A*, calls at port *B* and then at port *C* and finally arrives at her destination *D*. The distance from *A* to *C* is $\frac{2}{3}$ of the total voyage and the distance from *B* to *D* is $\frac{3}{4}$ of the total voyage. What fraction of the voyage is the distance from *B* to *C*?

Multiplication of Fractions.

The product of two or more fractions has for its numerator the product of the numerators, and for its denominator the product of the denominators, reduced to their lowest terms.

Factors common to any of the numerators and denominators may be divided into both to simplify the multiplication.

Mixed numbers must always be changed into improper fractions before multiplying.

Example. Find the product of $1\frac{9}{16}\times 4\frac{4}{5}\times 1\frac{2}{3}$.

$$\text{Product}=\frac{25}{\underset{2}{\cancel{16}}}\times\frac{\overset{8}{\cancel{24}}}{\cancel{5}}\times\frac{\cancel{5}}{\cancel{3}}=\frac{25}{2}=12\tfrac{1}{2}.$$

Exercise XII.

1. Find the value of:

(a) $\frac{2}{3}\times\frac{9}{11}$. (b) $\frac{3}{4}\times\frac{8}{9}$. (c) $\frac{7}{12}\times\frac{3}{14}$. (d) $\frac{5}{9}\times\frac{12}{25}$.
(e) $\frac{15}{16}\times\frac{20}{21}$. (f) $\frac{33}{40}\times\frac{72}{121}$. (g) $\frac{51}{60}\times\frac{5}{68}$. (h) $\frac{39}{40}\times\frac{32}{65}$.

2. Find the value of:

(a) $1\frac{1}{4}\times\frac{2}{3}$. (b) $\frac{3}{5}$ of $2\frac{1}{2}$. (c) $2\frac{1}{4}\times 1\frac{1}{3}$. (d) $3\frac{3}{8}\times 3\frac{1}{3}$.
(e) $5\frac{1}{3}\times 1\frac{1}{4}$. (f) $11\frac{4}{7}\times 5\frac{4}{9}$. (g) $1\frac{11}{25}$ of $1\frac{19}{36}$. (h) $2\frac{5}{12}\times 1\frac{7}{29}$.

3. Simplify:

(a) $\frac{9}{10}\times\frac{5}{8}\times\frac{4}{9}$. (b) $5\frac{1}{4}\times\frac{5}{7}\times 2\frac{2}{5}$. (c) $\frac{7}{12}$ of $9\frac{1}{3}\times 3\frac{3}{5}$.
(d) $2\frac{13}{16}\times 1\frac{1}{3}\times 1\frac{5}{9}$. (e) $2\frac{1}{10}\times\frac{5}{7}$ of 24. (f) $3\frac{3}{4}\times 1\frac{7}{8}\times 4\frac{4}{5}$.

4. Distance run = Speed × Time taken.

If the speed is in knots and the time in hours, or fractions of an hour, the distance run is in nautical miles.

Find the distance run under the following conditions:

	Speed in knots	Time in hours	Distance in N.M. (to nearest N.M.)
(a)	$10\frac{1}{2}$	$5\frac{1}{4}$	
(b)	$15\frac{3}{4}$	$11\frac{1}{2}$	
(c)	$18\frac{1}{2}$	$12\frac{2}{3}$	
(d)	$21\frac{1}{4}$	$9\frac{4}{5}$	

Problems involving Multiplication of Fractions.

Exercise XIII.

1. A ship's clock gains $1\frac{3}{5}$ sec. per day. How much will it gain in the following times?

(a) 5 days. (b) $6\frac{1}{4}$ days. (c) $3\frac{3}{4}$ days.

2. A ship's engines require $1\frac{3}{4}$ lb. of coal every hour for each horse-power developed. If the engines develop 2800 H.P., what is the consumption of coal per day of 24 hr.? (Answer in tons.)

3. A cruiser's measurements are: length 636 ft., breadth 75 ft., depth 33 ft. What are the corresponding measurements in inches of a model of scale $\frac{1}{12}$ in. = 1 ft.?

4. Two of the various kinds of thermometer used for measuring temperature are the Fahrenheit and the Centigrade thermometers. On the Fahrenheit thermometer the freezing point of water is 32° F. and the boiling point of water 212° F. (a difference of 180° F.). On the Centigrade thermometer the freezing point of water is 0° C. and the boiling point 100° C. (a difference of 100° C.).

To convert °C. into °F. first multiply the number of °C. by $\frac{180}{100}$, i.e. $\frac{9}{5}$, and add 32.

EXAMPLE 1. Change 60° C. into °F.

$$60° \text{ C.} = (60 \times \tfrac{9}{5}) + 32° \text{ F.}$$
$$= 108 + 32 = 140° \text{ F.}$$

To convert °F. into °C. first subtract 32 from the number of °F. and multiply by $\frac{5}{9}$.

EXAMPLE 2. Convert 104° F. into °C.

$$104° \text{ F.} = (104 - 32) \times \tfrac{5}{9}° \text{ C.}$$
$$= 72 \times \tfrac{5}{9} = 40° \text{ C.}$$

Convert these °C. into °F.:

(*a*) 35° C. (*b*) 17° C. (*c*) $62\frac{1}{2}$° C. (*d*) −25° C.

Convert these °F. into °C.:

(*e*) 158° F. (*f*) 185° F. (*g*) 71° F. (*h*) 14° F.

5. In an aerial battle $\frac{1}{10}$ of the total attacking force is destroyed in the first encounter and $\frac{1}{9}$ of the remainder in a second engagement. If 16 machines return, how many started out?

6. Two coal bunkers on a merchant ship are of equal capacity. From the first bunker $\frac{2}{3}$ of its total capacity has been used and, from the second, $\frac{2}{5}$.

(*a*) How much coal must be used from the bunker containing the larger quantity to equalise the amounts in each bunker?

(*b*) How much coal would have to be taken from the bunker with the larger quantity and transferred to the lesser quantity to equalise the amounts in each bunker? Answer in each case as a fraction of a bunker.

7. A vessel burning oil fuel uses $\frac{1}{4}$ of its tank capacity while steaming north and $\frac{1}{5}$ of the remainder while steaming east. What fraction of the tank capacity remained?

If the tank still contained 60 tons of oil, what was the weight of the original supply?

8. From a coil of hemp, $\frac{2}{5}$ of the coil has been issued for hammock lashings and $\frac{1}{3}$ of the coil for lanyards. There still remain 32 fathoms in the coil. What was the full length of the coil?

9. A ship's crew consists of equal numbers in both the port and starboard watches. $\frac{1}{4}$ of the port watch and $\frac{5}{12}$ of the starboard watch are on leave.

(*a*) What fraction of the ship's crew is on leave?

(*b*) If the number of men left aboard is 32, how many are there in the crew?

10. At half tide the depth of water at a certain place was 18 ft. When the tide was $\frac{2}{3}$ full the depth of water was 20 ft. Find the depth at (*a*) high water, (*b*) low water.

Division of Fractions.

As in multiplying, always change mixed numbers to improper fractions. Then invert the divisor, and multiply.

EXAMPLE. Simplify $2\frac{3}{5} \div \frac{11}{15}$.

$$2\frac{3}{5} \div \frac{11}{15} = \frac{13}{5} \div \frac{11}{15} = \frac{13}{\cancel{5}} \times \frac{\overset{3}{\cancel{15}}}{11} = \frac{39}{11} = 3\frac{6}{11}.$$

EXERCISE XIV.

1. Find the value of:

(*a*) $\frac{7}{8} \div 5$. (*b*) $25 \div \frac{5}{8}$. (*c*) $\frac{9}{14} \div \frac{18}{35}$. (*d*) $\frac{3}{7} \div \frac{9}{49}$.

(*e*) $33\frac{1}{3} \div 6\frac{1}{4}$. (*f*) $18\frac{3}{4} \div 3\frac{1}{8}$. (*g*) $23\frac{4}{5} \div 4\frac{1}{4}$. (*h*) $8\frac{1}{10} \div 4\frac{1}{2}$.

(*i*) $21\frac{7}{11} \div \frac{17}{132}$. (*j*) $1\frac{5}{39} \div \frac{11}{65}$.

2. Speed = Distance travelled ÷ Time taken.

If the distance is in nautical miles and the time in hours, the speed is in knots. Calculate the speeds in the following cases:

	Distance in N.M.	Time in hours	Speed in knots
(*a*)	$189\frac{5}{8}$	$10\frac{1}{4}$	
(*b*)	$134\frac{1}{3}$	$8\frac{2}{3}$	
(*c*)	$209\frac{3}{8}$	$16\frac{3}{4}$	
(*d*)	165	$7\frac{1}{3}$	

3. Time taken = Distance travelled ÷ Speed.

If the distance is in nautical miles and the speed in knots, the time is in hours (or fractions of an hour). Find the time taken in the following cases:

	Distance in N.M.	Speed in knots	Time in hours
(*a*)	184	16	
(*b*)	$252\frac{15}{16}$	$14\frac{1}{4}$	
(*c*)	$129\frac{1}{6}$	$12\frac{1}{2}$	
(*d*)	$216\frac{1}{3}$	11	

4. Area of a rectangle (or oblong) = Length × Breadth, i.e. Breadth = Area ÷ Length.

A photographic plate has an area of $13\frac{13}{16}$ sq. in. Its length is $4\frac{1}{4}$ in. What is its breadth?

Problems involving Division of Fractions.

EXERCISE XV.

1. Sizes of wire stays are expressed in terms of their circumferences, and it is sometimes required to know the diameters of such wires.

We know that $\quad \text{Circumference} = \frac{22}{7} \times \text{Diameter},$

i.e. $\quad \text{Diameter} = \text{Circumference} \div \frac{22}{7}.$

What are the diameters of the following wire stays?

(*a*) $3\frac{1}{2}$ in. (*b*) $1\frac{1}{4}$ in. (*c*) $2\frac{1}{4}$ in. (*d*) $2\frac{3}{4}$ in.

2. A ship's clock was 20 sec. slow at noon on 10 March and $8\frac{4}{5}$ sec. slow at noon on 18 March. What was the daily rate of the clock, gaining or losing?

3. A ship's clock was $13\frac{4}{5}$ sec. fast at 8.30 p.m. on 5 April and $20\frac{1}{5}$ sec. fast at 9.30 p.m. on 8 April. Find the daily rate of the clock, gaining or losing, to the nearest $\frac{1}{5}$ sec.

4. If a watch has a daily rate of $2\frac{4}{5}$ sec. gaining and at noon on 1 June it is $29\frac{2}{5}$ sec. slow, when will it show correct time?

5. There is a type of fractional problem in which the answer appears to be wrong, unless the process is clearly understood.

EXAMPLE. How many pieces of string $3\frac{1}{2}$ in. long can be cut from a length $7\frac{1}{4}$ in. and how much remains?

The answer to this is obviously 2, with $\frac{1}{4}$ in. left over.

By working with division of fractions, we have

$$\frac{7\frac{1}{4}}{3\frac{1}{2}} = \frac{29}{4} \div \frac{7}{2} = \frac{29}{\underset{2}{\cancel{4}}} \times \frac{\cancel{2}}{7} = \frac{29}{14} = 2\frac{1}{14}.$$

The fraction $\frac{1}{14}$ does *not* mean that $\frac{1}{14}$ in. is left, but $\frac{1}{14}$ of the *length* that is being cut off, i.e. $\frac{1}{14}$ of $3\frac{1}{2}$ in. $= \frac{1}{\underset{2}{\cancel{14}}} \times \frac{7}{2}$ in. $= \frac{1}{4}$ in.

How many lengths of rope each $4\frac{5}{8}$ fathoms can be cut from a coil containing $32\frac{1}{2}$ fathoms and what length remains?

6. How many pieces of steel rod each $3\frac{1}{2}$ in. long can be cut from a length 3 ft. 8 in. and how much will be left over? Allow in each piece $\frac{1}{16}$ in. wastage due to saw-cut.

7. If a cubic foot of fresh water weighs $62\frac{1}{2}$ lb., how many cubic feet are there in one English ton of fresh water?

8. How many cubic feet of fresh water are there in one metric ton of 2204 lb.?

9. If crude petroleum weighs $55\frac{1}{4}$ lb. per cubic foot and 1 cu. ft. is equivalent to $6\frac{1}{4}$ gallons, find the weight of crude petroleum in lb. per gallon.

10. A vessel of draught $18\frac{1}{2}$ ft. (i.e. the bottom of the ship's keel is $18\frac{1}{2}$ ft. below the surface of the water) is floating in $27\frac{3}{4}$ fathoms of water. What multiple of the draught is the depth of water?

11. A ship's bilge pump can pump $112\frac{1}{2}$ gallons in $3\frac{3}{4}$ min. What is the "capacity" of the pump in gallons per minute?

12. A seaplane has a petrol tank capacity of 350 gallons, of which $\frac{1}{7}$ must be kept as reserve. If the aircraft uses 40 gallons per hour while in flight, what is her possible flying time (endurance)?

Fractions of Quantities.

Any fraction of a quantity is itself a quantity. Thus a quarter of an hour is not merely a number, it is a quarter of 60 min. = 15 min.

EXAMPLE 1. What is the value of $2\frac{7}{8}$ of 2 tons 8 cwt.?

$$2 \text{ tons } 8 \text{ cwt.} = 2\tfrac{8}{20} \text{ tons} = 2\tfrac{2}{5} \text{ tons.}$$

Therefore

$$2\tfrac{7}{8} \text{ of } 2 \text{ tons } 8 \text{ cwt.} = 2\tfrac{7}{8} \times 2\tfrac{2}{5} \text{ tons}$$

$$= \frac{23}{\cancel{8}_{\,2}} \times \frac{\cancel{12}^{\,3}}{5} = \frac{69}{10} \text{ tons} = 6\tfrac{9}{10} \text{ tons,}$$

i.e. 6 tons 18 cwt.

EXAMPLE 2. What fraction of 1 hour are $2\frac{1}{2}$ minutes?

Rule. Place the quantity immediately following the word "of" in the denominator, thus: $\dfrac{\quad}{1 \text{ hour}}$.

Place the other quantity in the numerator, thus: $\dfrac{2\frac{1}{2} \text{ minutes}}{1 \text{ hour}}$.

Bring both the numerator and denominator to the same units, i.e. minutes:

$$\frac{2\frac{1}{2} \text{ minutes}}{1 \text{ hour}} = \frac{\cancel{5}}{2} \times \frac{1}{\cancel{60}_{\,12}} = \frac{1}{24}.$$

EXAMPLE 3. The scale of a map or chart is often stated as a fraction, called the *Representative Fraction* (R.F.) of the map.

Suppose, upon the plan of a harbour, two mooring buoys are represented and measure 3 inches apart. These two buoys are known to be actually 2 nautical miles apart.

The scale of the plan is, therefore, 3 inches = 2 nautical miles. If we assume the nautical mile to be 6080 feet, we have:

$$\text{Representative Fraction} = \frac{\text{Distance on plan}}{\text{Distance on "ground"}} = \frac{3 \text{ in.}}{2 \text{ N.M.}}$$

$$= \frac{\cancel{3} \text{ in.}}{2 \times 6080 \times \underset{4}{\cancel{12}} \text{ in.}} = \frac{1}{48{,}640}.$$

In other words, on this plan:

A map distance of 1 in. represents 48,640 in. on the ground.

„ 1 ft. „ 48,640 ft. „

„ $3\frac{1}{2}$ in. „ $3\frac{1}{2} \times 48{,}640$ in. „

EXERCISE XVI.

1. What fraction is
   (a) $12\frac{1}{2}$ min. of 1 hour? (b) $2\frac{1}{4}$ gal. of 30 gal.?
   (c) $80\frac{1}{2}$ miles of 110 miles? (d) 1 min. 35 sec. of 4 min. 20 sec.?
   (e) 2 min. 5 sec. of $8\frac{1}{2}$ min.? (f) 1 inch of 1 mile?

2. Find the value of
   (a) $2\frac{3}{4}$ of $2\frac{2}{3}$ fathoms. (b) $\frac{5}{8}$ of 12 tons 16 cwt.
   (c) $\frac{4}{7}$ of $164\frac{1}{2}$ gal. (d) $\frac{7}{9}$ of £1. 8*s*. 6*d*.
   (e) $2\frac{1}{4}$ of 324 feet, in fathoms. (f) $5\frac{1}{3}$ of $5\frac{1}{4}$ guineas.

3. Find the R.F. in each of the following cases:

	Map distance	Ground distance
(a)	6 in.	$1\frac{1}{2}$ statute miles
(b)	$1\frac{3}{4}$ in.	$\frac{7}{8}$ nautical mile
(c)	$\frac{1}{4}$ in.	2 cables

4. Express these R.F.S as statute miles per inch:

   (a) $\frac{1}{21{,}120}$. (b) $\frac{1}{158{,}400}$. (c) $\frac{1}{88{,}704}$.

5. On a chart, of R.F. $\frac{1}{109{,}440}$, a channel entry between a port and starboard buoy measures $1\frac{1}{2}$ in. What is the width of this channel in nautical miles?

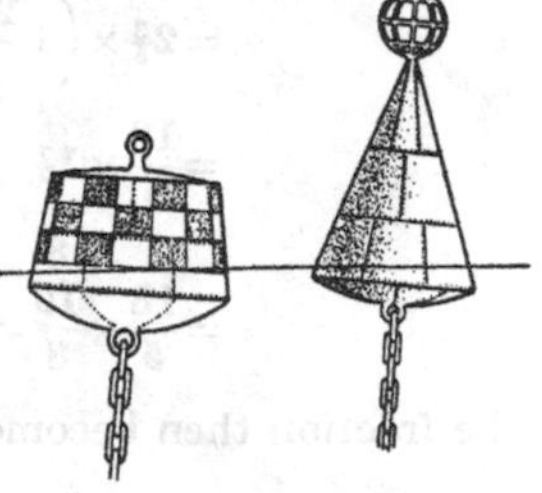

Port-hand Can Buoy   Starboard-hand Conical Buoy

6. What fraction of a nautical mile (6080 ft.) is a statute mile?

7. On a ruler 10 in. are equal in length to $25\frac{2}{5}$ cm. What fraction is (a) 1 cm. of 1 in., (b) 1 in. of 1 cm.?

8. A vessel steaming at 12 knots increases her speed by $\frac{1}{5}$. What is her new speed? If later she reduces this new speed by $\frac{1}{4}$, what is her final speed?

9. A ship steaming at 15 knots reduces speed by $\frac{1}{4}$. What is her new speed? If she then increases this new speed by $\frac{1}{4}$, what is her final speed?

10. By what fraction of the new speed in question 9 would the vessel have had to increase to return to her original speed of 15 knots?

11. A ship approaching land from seaward takes soundings and finds the depth of water to decrease at the rate of 3 fathoms in 2 nautical miles. If 1 N.M. = 2000 yd., express this "shoaling rate" as a fraction.

12. Above a bar at the entrance to a river the depth at low water on a certain day is found to be $3\frac{1}{2}$ fathoms. At high water the tide is found to have risen 15 ft. 3 in. above low water. What fraction is the low-water depth above the bar of the high-water depth?

Complex Fractions.

Complex fractions are those which may be made up of fractional numerators and fractional denominators and may include addition, subtraction, multiplication and division of fractions to be worked simultaneously. These must be simplified in accordance with the "Order of Signs" as shown in the following table:

(*a*) Brackets must be cleared first.
(*b*) Numbers connected by "of" next.
(*c*) $\times$ and $\div$ signs are equal in value and are taken in their order of occurrence.
(*d*) $+$ and $-$ signs are also equal in value and come last, in order of occurrence.

It is best to simplify a complex fraction by working the numerator and denominator separately.

EXAMPLE. Simplify $\dfrac{2\frac{3}{5}\times(1\frac{1}{4}+\frac{5}{8})}{1\frac{2}{5}\div(\frac{5}{6}-\frac{3}{4})}$.

Numerator

$$2\tfrac{3}{5}\times(1\tfrac{1}{4}+\tfrac{5}{8})$$
$$=2\tfrac{3}{5}\times\left(1\,\frac{2+5}{8}\right)$$
$$=\frac{13}{5}\times1\tfrac{7}{8}$$
$$=\frac{13}{\cancel{5}}\times\frac{\overset{3}{\cancel{15}}}{8}=\frac{39}{8}.$$

Denominator

$$1\tfrac{2}{5}\div(\tfrac{5}{6}-\tfrac{3}{4})$$
$$=1\tfrac{2}{5}\div\left(\frac{10-9}{12}\right)$$
$$=\frac{7}{5}\div\frac{1}{12}$$
$$=\frac{7}{5}\times\frac{12}{1}=\frac{84}{5}.$$

The fraction then becomes

$$\frac{\frac{39}{8}}{\frac{84}{5}}=\frac{\overset{13}{\cancel{39}}}{8}\times\frac{5}{\underset{28}{\cancel{84}}}=\frac{65}{224}.$$

Exercise XVII.

Simplify the following:

1. $\frac{1}{2}+\frac{2}{3}$ of $\frac{2}{5}+\frac{3}{4}$. 2. $(\frac{1}{2}+\frac{2}{3})$ of $\frac{2}{5}+\frac{3}{4}$. 3. $(\frac{1}{2}+\frac{2}{3})$ of $(\frac{2}{5}+\frac{3}{4})$.

4. $\frac{1}{2}+\frac{2}{3}$ of $(\frac{2}{5}+\frac{3}{4})$. 5. $\frac{3}{4}\div 1\frac{1}{2}\times 1\frac{1}{3}$. 6. $\frac{3}{4}\div 1\frac{1}{2}$ of $1\frac{1}{3}$.

7. $1\frac{1}{2}+1\frac{2}{3}-\frac{1}{5}$ of $1\frac{7}{8}-(\frac{3}{4}-\frac{5}{8})$.

8. (*a*) $2\frac{1}{4}\div 1\frac{2}{3}\times 3\frac{3}{4}$. (*b*) $\dfrac{2\frac{2}{5}\times(\frac{3}{4}+\frac{7}{8})}{1\frac{1}{4}\div(\frac{5}{6}-\frac{1}{4})}$.

9. (*a*) $3\frac{1}{2}\div 1\frac{3}{4}\times 1\frac{1}{8}$. (*b*) $\dfrac{1\frac{1}{2}\text{ of }(1\frac{1}{2}+\frac{2}{3})}{1\frac{1}{2}\div(1\frac{1}{2}+\frac{2}{3})}$.

10. (*a*) $\frac{5}{8}\div 3\frac{3}{4}\times 3\frac{2}{3}$. (*b*) $\dfrac{1\frac{1}{2}+\frac{2}{3}\text{ of }2\frac{1}{2}}{(1\frac{1}{2}+\frac{2}{3})\text{ of }2\frac{1}{2}}$.

11. (*a*) $\frac{5}{9}\times 2\frac{1}{7}\times\frac{7}{25}$. (*b*) $\dfrac{\frac{3}{4}+1\frac{1}{2}\times 1\frac{1}{3}}{(\frac{2}{3}+1\frac{1}{4})\times 2\frac{1}{2}}$.

12. (*a*) $3\frac{1}{2}\div 1\frac{3}{4}+2\frac{1}{8}$. (*b*) $\dfrac{\frac{3}{4}+\frac{2}{5}}{\frac{3}{4}-\frac{2}{5}}\div\dfrac{\frac{2}{3}+\frac{3}{7}}{1\frac{1}{3}}$.

Miscellaneous Fractional Problems.

Exercise XVIII.

1. At low water on a certain day a tide pole registered 4 ft. of water. At the next high water the tide had risen 11 ft. and $\frac{3}{4}$ of the pole was then submerged. How many feet of the pole were visible above high water?

2. After reducing speed by $\frac{1}{7}$ a ship steamed a distance of $56\frac{1}{4}$ N.M. in $3\frac{3}{4}$ hr. What was the original speed of the ship in knots?

3. A ship's engines are developing 3680 indicated horse power (I.H.P.) and are driving the ship at a speed of 18 knots. If her engines consume $1\frac{3}{4}$ lb. of coal for each I.H.P. every hour, find the decrease in tons' weight of the ship due to fuel consumption during a run of 621 N.M.

4. A ship discharges $\frac{1}{12}$ of her cargo at her first port of call and $\frac{3}{11}$ of the remainder at her second. If the cargo then remaining weighs 1200 tons, what was the weight of the initial cargo?

5. At noon on 10 January a ship's clock was $10\frac{2}{5}$ sec. slow by Greenwich Mean Time. If its daily rate was $1\frac{3}{5}$ sec. losing, what was the error of the clock at 9 p.m. on 15 January?

6. A ship's clock was $4\frac{4}{5}$ sec. slow at noon on 8 May. Its daily rate was $2\frac{4}{5}$ sec. gaining. What was the time by the ship's clock when Greenwich Mean Time was 10 p.m. on 14 May?

7. A vessel uses $\frac{1}{5}$ of its total fuel tonnage in reaching its first port of call and $\frac{2}{5}$ of the remainder between the first and second ports of call. If she then has a fuel tonnage of 360 tons, what was her original fuel tonnage?

8. A ship has completed half her run from Gibraltar to Southampton. After steaming a further 179 N.M. she still has $\frac{1}{3}$ of the total distance left to go. What is the distance from Gibraltar to Southampton?

9. One gallon of a mixture of turpentine and oil consists of 1 part turpentine and 5 parts oil. To mix a particular paint another quart of oil is added to the above gallon. What fraction of the resulting mixture is turpentine?

10. A deck watch was $13\frac{4}{5}$ sec. fast by the chronometer at 6 a.m. on 3 April. If its daily rate was $1\frac{4}{5}$ sec. losing, at what time and on what date did the two watches show the correct time?

11. The depth of water at a certain place at low water spring tide (L.W.S.T.) is known to be 3 fathoms. A ship of draught 20 ft. passes over this place when the tide is 10 ft. 6 in. above L.W.S.T. What fraction of the depth at L.W.S.T. is then below the ship's keel?

12. An oil fuel bunker contained $\frac{1}{10}$ of its fuel capacity. 154 tons of oil were then pumped in and it was then $\frac{4}{5}$ full. How many tons of oil did the bunker then contain and what was its full capacity in tons?

13. A certain pump can deliver $52\frac{1}{2}$ tons of fuel oil in $3\frac{1}{4}$ min. How long will this pump take to load a fuel tank with 330 tons?

14. Two vessels, *A* and *B*, leave the same harbour at the same time, steaming in opposite directions. The speed of *A* is $\frac{4}{5}$ that of *B*. After $3\frac{1}{2}$ hr. they are 126 N.M. apart. What are the speeds of *A* and *B*?

15. By chronometer time at 8 a.m. on 26 March a deck watch was $9\frac{3}{5}$ sec. slow. If its daily rate was $2\frac{2}{5}$ sec. gaining, when, and on what date, was the watch $9\frac{3}{5}$ sec. fast by chronometer time?

16. Two ships, *A* and *B*, leave the same harbour at the same time, steaming in the same direction. The speed of *A* is $\frac{5}{7}$ that of *B*. If, after $4\frac{1}{2}$ hr., they are $22\frac{1}{2}$ N.M. apart, what are the speeds of *A* and *B*?

17. A lump of a certain alloy weighs 3 lb. 15 oz. and contains $2\frac{1}{4}$ lb. of copper, 1 lb. 5 oz. of zinc and 6 oz. of tin. What would be the weight of copper in one ton of this alloy?

18. A ship steaming with the tide travelled $1\frac{1}{2}$ times as fast as when steaming against the tide. When steaming with the tide she covered $2\frac{1}{2}$ N.M. in $12\frac{1}{2}$ min. What was her speed in knots against the tide?

19. A sea-plane starts out with a petrol load of 360 gallons and uses $\frac{7}{16}$ of it during a flight. If petrol weighs $7\frac{1}{5}$ lb. per gallon, what weight of petrol still remains?

20. A ship's company consists of equal numbers in both the Port and Starboard watches. $\frac{1}{6}$ of the Port watch and $\frac{3}{4}$ of the Starboard watch are on leave. (*a*) What fraction of the ship's company is still on board? (*b*) If the number of men on leave is 66, what is the strength of the ship's company?

Pumps and Leaks.

Consider now a problem such as the pumping of water from a flooded compartment of a ship, which has been holed, and the time that will be taken to do this, with the leak still operating.

EXAMPLE. A ship's pump can empty a flooded compartment in 5 hours, when watertight. The compartment becomes holed and the leak, acting alone, floods the compartment in $8\frac{1}{2}$ hours. If the compartment is half full when the pump is turned on, how long will it take to empty it?

The pump, with no leak, empties the compartment in 5 hours.

Therefore the pump, with no leak, empties $\frac{1}{5}$ of the compartment in 1 hour.

The leak, acting alone, floods the compartment in $8\frac{1}{2}$ hours $=\frac{17}{2}$ hours.

Therefore the leak will flood $\frac{2}{17}$ of the compartment in 1 hour.

Thus, when both are operating, the pump will remove $\frac{1}{5}-\frac{2}{17}$ of the compartment in 1 hour

$$=\frac{17-10}{85}=\frac{7}{85}$$

of the compartment in 1 hour.

The pump, with the leak operating, would therefore empty the full compartment in $\frac{1}{\frac{7}{85}}$ hours $=\frac{85}{7}$ hours.

But the compartment is only half full when the pump is turned on, so that the time taken to empty it would be

$$\tfrac{1}{2} \text{ of } \tfrac{85}{7}=\tfrac{85}{14} \text{ hours}=6\tfrac{1}{14} \text{ hours},$$

i.e. 6 hours 4 minutes (to the nearest minute).

EXERCISE XIX.

1. One pump acting alone can empty a fully flooded compartment in $5\frac{1}{2}$ hr. Another pump acting alone can empty the same compartment in $3\frac{3}{4}$ hr. If both pumps start operating on the flooded compartment at noon, when will it be empty? (Answer to nearest minute.)

2. A compartment that is holed floods completely in $2\frac{1}{2}$ hr. A pump can empty this full compartment in $1\frac{3}{4}$ hr. if there is no leak. If fully flooded when the pump starts operating, how long will it take to clear the compartment with the leak still acting?

3. Two pumps are capable of discharging 200 and 350 gallons per hour respectively. They are started up together to empty a fully loaded oil storage tank. After $2\frac{1}{2}$ hr. it is found that $\frac{5}{16}$ of the oil has been removed. Find (*a*) the capacity of the tank in gallons, (*b*) how many gallons of oil remain after $5\frac{1}{2}$ hr. pumping.

4. A compartment is holed and floods completely in $6\frac{1}{4}$ hr. Two pumps are capable of emptying the fully flooded compartment in $2\frac{1}{2}$ hr. and 3 hr. respectively if there is no leak. How long would it take to empty the compartment, with the leak still acting, if both pumps are set in operation when the compartment is $\frac{3}{8}$ full?

5. A rectangular compartment, below the water line, becomes holed at 10 a.m. The compartment is 8 ft. high and at 10.30 a.m. the lead-line shows the water to have risen to a depth of 4 ft. 6 in. in the compartment. At 10.30 a.m. clearance pumps are started, but at 10.45 a.m. the water has gained another $3\frac{1}{2}$ in. in spite of the pumping. Assuming the rate of the leak to be constant, at what time will the compartment be full, with the pumps acting?

6. One pump working alone can empty a fully flooded compartment in $4\frac{1}{2}$ hr. A second pump has a discharge rate of 600 gallons per hour. Both pumps operating together can empty the compartment in 2 hr. Find (*a*) the number of gallons per hour discharged by the first pump, (*b*) the capacity of the compartment in gallons.

7. At 10 a.m. a compartment below the water line is holed, and by 11.30 a.m. the compartment is $\frac{3}{4}$ full. Two pumps A and B are started up at 11.30 a.m. and by 1.30 p.m. the compartment is only $\frac{1}{3}$ full. At 1.30 p.m. pump A breaks down and is out of action for 30 min., during which time the compartment has become $\frac{1}{2}$ full. What fraction of the compartment volume would each pump, acting alone, empty per hour if the compartment was fully flooded and the leak stopped?

8. A compartment below the water line of a vessel has a capacity of 1800 cu. ft. After being holed the compartment floods to $\frac{1}{5}$ of its capacity in 1 hour. During a second hour the water increases to $\frac{7}{30}$ of the capacity in spite of a pump which has been working for the second hour. If 1 cu. ft. $= 6\frac{1}{4}$ gallons, find the discharge rate in gallons per hour of an additional pump that will just maintain the water level in the compartment, without further rise.

DECIMALS

Consider, first, the meaning of the number 243.

This, as we know, means the sum of 2 hundreds + 4 tens + 3 units.

Suppose that we wish to write the sum of 2 tenths + 4 hundredths + 3 thousandths.

This is written

$$\frac{2}{10}+\frac{4}{100}+\frac{3}{1000}=\frac{200+40+3}{1000}=\frac{243}{1000}$$

when expressed as a vulgar fraction.

It may also be expressed in the form of a *decimal fraction* thus:

$$\cdot 243.$$

The first figure after the decimal point denotes the number of tenths, the second figure the number of hundredths and the third figure the number of thousandths.

Suppose, now, that we wish to write the number two hundred and forty in figures. This is 240.

The 0 at the end denotes that there are no units in this number, but the 0 must be included to show that this is so and to show that the next figure, 4, is the number of tens.

In the same way, if we wish to write four hundredths and three thousandths as a decimal fraction, we write ·043. In this case the 0 following the decimal point shows that there are no tenths and also shows that the next figure, 4, is the number of hundredths.

So that if we wish to write, as a decimal, the number two hundreds, four tens, four hundredths and three thousandths, we should write: 240·043.

We know that it does not alter the value of a whole number to put one or more 0's in front of it.

Thus 00243 has the same value as 243.

Similarly ·24300 has the same value as ·243, since the final 0's mean simply that there are no tenthousandths and no hundredthousandths.

So that 00243·24300 is the same as 243·243.

Significant Figures.

The answer to a decimal problem is often asked for to a certain number of *significant figures*.

Although the number of 0's immediately following the decimal point have a definite meaning, they do not, without the rest of the figures, give any indication of the actual value of the decimal.

For example, in the decimal ·00243 the two 0's denote that there are no tenths and no hundredths. It is the figures 243 that tell us the value of the decimal and these are therefore called *significant figures*.

The decimal ·00243 has, therefore, three significant figures.

It is usual, for the sake of clarity and to avoid risk of error, to insert a 0 before the decimal point of any decimal with no whole number, e.g. ·243 is always written 0·243, and ·00243 is written 0·00243.

Addition and Subtraction of Decimals.

When adding or subtracting decimals the points must be kept always under one another.

Thus 12·88 + 1·096 + 0·87. And 241·091 − 12·87.

```
 12·88          241·091
  1·096          12·87
  0·87          -------
 ------         228·221
 14·846
```

Exercise XX.

1. Add:

(*a*) 0·568 + 19·38 + 1·163 + 0·2218.

(*b*) 30·157 + 5·04 + 0·786 + 19·2.

(*c*) 2·0508 + 0·77 + 70·908 + 11·8903.

(*d*) 87·5 + 8·75 + 0·875 + 0·0875 + 0·00875.

(*e*) 1·98 + 18·694 + 246·224 + 0·4 + 17·935.

(*f*) 0·05 + 0·734 + 0·9049 + 0·0007 + 0·0616.

(*g*) 126·67 + 3·4716 + 10 + 81·2001 + 69·749 + 8·8 + 80·08.

(*h*) 192·8 + 17·96 + 2·3 + 5·841 + 121·3 + 64·87.

(*i*) 0·415 + 2·896 + 39·771 + 15·875 + 21·81.

2. Subtract:

(*a*) 21·08 − 14·89. (*b*) 13·34 − 8·891. (*c*) 12·214 − 7·006.

(*d*) 9 − 3·876. (*e*) 5·302 − 3·479. (*f*) 4·09 − 3·8654.

(*g*) 98·65 − 9·752. (*h*) 40·47 − 4·047. (*i*) 0·212121 − 0·12121.

3. The normal atmospheric pressure for the British Isles is 29·92 in. of mercury. If on a certain day the mercury barometer reads 31·11 in., how much is this above normal?

4. The lowest recorded barometer reading for the British Isles is 27·33 in. How many inches is this less than normal?

5. In weather forecasting (or meteorology) atmospheric pressure is expressed in millibars and the normal reading for the British Isles is 1013·2 millibars (mb.). State the pressure in millibars when it is (*a*) 32·6 mb. above normal, (*b*) 82·2 mb. below normal.

6. Rainfall is measured in inches. These are the monthly totals for a certain year at Kirkwall:

4·19; 3·55; 3·45; 1·96; 1·76; 1·93; 2·9; 3·16; 3·23; 4·74; 4·57; 4·64.

Find (*a*) the total annual rainfall, (*b*) difference between the greatest and least month's rain.

Multiplication of Decimals.

Example. Multiply 12·42 by 3·6.

First multiply without points:

```
 1242
   36
 ----
 7452
3726
-----
44712
```

Count the total number of figures *after* the decimal points in the two numbers to be multiplied. In this case, 3.

Count 3 figures from the right in the product and insert the decimal point, i.e.

The answer is 44·712.

Exercise XXI.

1. Find the product of:

(*a*) 4·24 × 1·4. (*b*) 21·4 × 0·013. (*c*) 216·2 × 2·2.
(*d*) 0·013 × 0·9. (*e*) 4·6 × 43·02. (*f*) 9·8 × 0·0803.
(*g*) 124·8 × 0·08. (*h*) 124·876 × 21·5. (*i*) 3·142 × 9·78.
(*j*) 0·7261 × 75·4.

2. Find the area, in square inches, of these rectangular plates (Area = Length × Breadth):

(*a*)		(*b*)		(*c*)	
Length	Breadth	Length	Breadth	Length	Breadth
3·25 in.	2·12 in.	4·25 in.	3·25 in.	7·63 in.	3·07 in.

3. If 1 millibar = 0·029 inch of mercury, convert the following millibar readings into barometric inches:

(*a*) 1007·2 mb. (*b*) 996·4 mb. (*c*) 1041 mb.

4. The temperature of the air usually decreases as altitude increases, and the average fall in temperature is 1·98° C. for each 1000 ft. rise in altitude. Normal sea-level temperature is taken to be 15° C. If the sea-level temperature is normal, what would be the temperature of the air at these altitudes?

(*a*) 4000 ft. (*b*) 5000 ft. (*c*) 10,000 ft. (*d*) 36,000 ft.

5. At low altitudes, atmospheric pressure falls approximately 1 millibar for every 30 feet rise in altitude. If the atmospheric pressure is normal at sea level (i.e. 1013·2 mb.), what are the readings at these altitudes?

(*a*) 6000 ft. (*b*) 7200 ft. (*c*) 12,000 ft.
(*d*) 5400 ft. (*e*) 3300 ft. (*f*) 1950 ft.

6. Find the altitudes at which the following readings were made, if sea-level pressure is normal:

(*a*) 780·2 mb. (*b*) 650 mb. (*c*) 926·4 mb.
(*d*) 900 mb. (*e*) 862·2 mb.

Division of Decimals.

Example 1. Divide 16·842 by 4·2.

First convert the divisor, 4·2, into a whole number by moving the decimal point one place to the right, i.e. 42.

Equalise this by moving the decimal point in the dividend, 16·842, one place to the right also, i.e. 168·42.

Insert the decimal point in the answer immediately after bringing down the first figure of the decimal in the dividend, thus

```
       4·01
   ________
42)168·42
   168
      42
      42
      ..
```

Example 2. Divide 151·262 by 3·61.

By moving the decimal point as required this becomes:

$$15126{\cdot}2 \div 361.$$

By division

```
        41·9
     ________
361)15126·2
    1444
      686
      361
      3252
      3249
      ...3
```

The division is not complete, as there is still a remainder.

But, since 15126·2 may be written 15126·20000...., we may bring down as many noughts as necessary and continue the division.

Should the answer be required to three decimal places, the division is carried on as follows:

```
        41·900
     __________
361)15126·2
    1444
      686
      361
      3252
      3249
      ...300
```

The answer to three decimal places is therefore 41·900.

If the *nearest* third decimal place is wanted, then the division must be carried to four places. If this fourth figure in the decimal is 5 or over, we add 1 to the third place.

In the above example the fourth figure of decimals is 8. Therefore the answer to the *nearest* third decimal place is 41·901.

Exercise XXII.

1. Simplify:

(*a*) 52·5 ÷ 2·5. (*b*) 28·08 ÷ 0·9. (*c*) 1·44 ÷ 1·2.
(*d*) 15·21 ÷ 1·17. (*e*) 0·0062 ÷ 2·5. (*f*) 183·82 ÷ 0·091.
(*g*) 264·708 ÷ 3·24. (*h*) 28·782 ÷ 3·69. (*i*) 91·008 ÷ 379·2.
(*j*) 0·4496 ÷ 11·24.

2. Express as a decimal to number of places stated:

(*a*) 0·51 ÷ 6·25 (to 3rd pl.). (*b*) 2·653 ÷ 7·4 (to 3rd pl.).
(*c*) 87 ÷ 426·8 (to 2nd pl.). (*d*) 12·85 ÷ 17·2 (to nearest 3rd pl.).
(*e*) 5·42913 ÷ 527·1 (to nearest 3rd pl.).
(*f*) 0·014 ÷ 2·3 (to 2nd pl.).
(*g*) 186·74 ÷ 53·2 (to nearest whole number or "integer" as it is called).
(*h*) 10·7358 ÷ 0·174 (to nearest integer).

3. The circumference of any circle is approximately equal to 3·14 times its diameter. What is the diameter of the earth to the nearest nautical mile if the length of the equator is 21,600 N.M.?

4. On a model globe 10° of the equator measure 1·22 in. Find the diameter of this globe to the nearest first decimal place.

5. The captain of a ship 5·7 N.M. from a signal gun on shore wishes to check his watch by the sound of the noon gun. If his watch shows 11 sec. past noon when he hears the gun, how much is his watch fast or slow? (Answer to the nearest 0·5 sec. Sound travels at 1090 ft. per sec. and 1 N.M. = 6080 ft.)

6. A ship in coastal waters, in a fog, sounds her foghorn. There is an echo of the foghorn's note reflected from high cliffs on the coast. If this echo is heard on the ship 27·6 sec. after the note is blown, how far off shore is the ship? (Sound travels at 1090 ft. per sec. and 1 N.M. = 6080 ft. Give your answer in nautical miles correct to the nearest second decimal place.)

Conversion of Decimals to Vulgar Fractions and Vice Versa.

Example 1. Express 2·465 as a fraction.

$$2{\cdot}465 = 2 + \tfrac{4}{10} + \tfrac{6}{100} + \tfrac{5}{1000}$$
$$= 2 + \frac{400+60+5}{1000} = 2\tfrac{465}{1000}.$$

The rule, therefore, is to place the decimal part as the numerator and for the denominator place a 1 for the decimal point and as many noughts as there are *useful* figures in the decimal.

This fraction should then be reduced to its lowest terms:

$$2\frac{\overset{93}{\cancel{465}}}{\underset{200}{\cancel{1000}}}=2\tfrac{93}{200}.$$

Note. The fraction equivalent to 4·031 is $4\frac{31}{1000}$, since the 0 following the decimal point is a useful figure.

EXAMPLE 2. Change $\frac{3}{8}$ to a decimal.

$\frac{3}{8}$ means 3 divided by 8. By simple division the answer is

0·375.

In the case of a fraction, in which the numerator is not exactly divisible by the denominator, the decimal is usually expressed to a stated number of places or significant figures.

EXAMPLE 3. Express $\frac{7}{9}$ as a decimal to the nearest third place.

By division, the answer is 0·778.

EXERCISE XXIII.

1. Convert the following decimals to vulgar fractions in their lowest terms:

(*a*) 0·5. (*b*) 0·95. (*c*) 0·15. (*d*) 0·94.
(*e*) 0·875. (*f*) 0·075. (*g*) 1·28. (*h*) 5·65.
(*i*) 2·45. (*j*) 1·025. (*k*) 2·0125. (*l*) 10·001.
(*m*) 3·504. (*n*) 10·375. (*o*) 0·0256. (*p*) 3·142.
(*q*) 6·0625. (*r*) 2·044. (*s*) 1·408. (*t*) 0·0011.

2. Convert into decimals:

(*a*) $\frac{1}{8}$. (*b*) $\frac{5}{8}$. (*c*) $\frac{3}{16}$. (*d*) $\frac{5}{16}$.
(*e*) $\frac{11}{16}$. (*f*) $\frac{9}{32}$. (*g*) $\frac{23}{32}$.

Convert into decimals (to nearest third decimal place):

(*h*) $\frac{6}{7}$. (*i*) $\frac{6}{13}$. (*j*) $\frac{2}{3}$. (*k*) $\frac{5}{9}$.
(*l*) $\frac{7}{17}$. (*m*) $\frac{11}{15}$. (*n*) $\frac{19}{230}$. (*o*) $3\frac{1}{7}$.

Decimal Values of Quantities.

EXAMPLE 1. A vessel has a maximum speed of 21 knots, but her most economical cruising speed is 17 knots. What decimal is her cruising speed of her maximum speed?

First express this as a fraction (see p. 24) thus:

$$\frac{\text{Cruising speed}}{\text{Maximum speed}}=\frac{17}{21}.$$

Converting this to a decimal by division, we have 0·809 to three decimal places.

EXAMPLE 2. What decimal is 2 ft. 3 in. of 1 fathom? Answer to the nearest second decimal place.

As a fraction, this is $\frac{2 \text{ ft. } 3 \text{ in.}}{1 \text{ fathom}} = \frac{2 \text{ ft. } 3 \text{ in.}}{6 \text{ ft.}}$.

Expressing both the numerator and denominator in the same units this becomes $\frac{2\cdot25 \text{ ft.}}{6 \text{ ft.}}$ or $\frac{27 \text{ in.}}{72 \text{ in.}}$, either of which, by division, is 0·375.

The answer, to the nearest second place, is therefore 0·38.

EXERCISE XXIV.

Give all the answers, unless otherwise stated, to the nearest third decimal place.

1. If one U.S. ton = 2000 English lb., what decimal is (*a*) a U.S. ton of an English ton, (*b*) an English ton of a U.S. ton?

2. If 136 English lb. = 125 Amsterdam lb., what decimal is (*a*) an English lb. of an Amsterdam lb., (*b*) an Amsterdam lb. of an English lb.?

3. The unit of length in China is a Ch'ih, where 1 ch'ih = 14·1 inches. What decimal is (*a*) a ch'ih of an English foot, (*b*) 1 English foot of 1 ch'ih?

4. Express £2·76 in pounds, shillings and pence to the nearest penny.

5. Express £4. 12*s*. 8*d*. as a decimal of a pound.

6. If a U.S. dollar is equal to 4*s*. 2*d*. in English money, what decimal is a U.S. dollar of £1?

7. The depth of water at a certain place at low water is 10 ft. 3 in. and at high water 23 ft. 8 in. What decimal is the low-water depth of the high-water depth?

8. What decimal is 1 yard of 1 metre, if 1 metre = 39·37 inches?

9. What decimal is 1 statute mile of 1 nautical mile (6080 feet)?

10. What decimal is 1° F. of 1° C.?

11. If 29·92 inches of mercury are equivalent to 1013·2 millibars atmospheric pressure, what decimal is 1 millibar of 1 "mercury" inch?

12. A ship's length is 627 ft. and her breadth 74·3 ft. What decimal is her length of her breadth?

Complex Decimals.

Priority must be given, as in complex fractions, to the "order of signs".

EXERCISE XXV.

Simplify, giving answers to three decimal places:

1. $4\cdot5 \div 2\cdot25 \times 10\cdot8$.

2. $4\cdot5 \div 2\cdot25$ of $10\cdot8$.

3. $(3\cdot1 + 2\cdot2 + 0\cdot16 + 1\cdot44) \times (1\cdot4 - 1\cdot38)$.

4. $(3 \cdot 6 - 1 \cdot 02 \times 2 \cdot 8) \div 0 \cdot 6$ of $1 \cdot 2$.

5. $12 \cdot 8 \div 3 \cdot 2 \times 1 \cdot 5 - 1 \cdot 75$.

6. $12 \cdot 8 \div 3 \cdot 2$ of $1 \cdot 5 - 1 \cdot 75$.

7. $9 \cdot 6 - 2 \cdot 4$ of $(2 \cdot 8 - 1 \cdot 35)$.

8. $\dfrac{2 \cdot 1 - 1 \cdot 5 \div (3 \times 0 \cdot 25)}{2 \cdot 1 + 1 \cdot 6 \div 0 \cdot 8 - 0 \cdot 1}$.

9. $\dfrac{2 \cdot 1 - 1 \cdot 2 \div 0 \cdot 3 \times 0 \cdot 25}{2 \cdot 1 - 1 \cdot 5 \div 0 \cdot 8 \text{ of } 2 \cdot 5}$.

10. $\dfrac{(1 \cdot 1 + 2 \cdot 2) - 0 \cdot 5\,(2 \cdot 2 - 1 \cdot 1)}{1 \cdot 1 - 2 \cdot 2\,(0 \cdot 5 \div 2 \cdot 5)}$.

Definitions of Terms Applied to Ships.

Every sea-going ship carries a certificate of registry on which are recorded the ship's dimensions. This record serves in part to identify the ship, and a Customs or Board of Trade officer is empowered to check the measurements if necessary.

Registered Length.

This, as recorded on the registry certificate, is the length of the ship from the foreside of the stem, at the extreme top, to the afterside of the sternpost.

Registered Breadth.

This is the ship's maximum width measured to the outside of the side plating.

Registered Depth.

This is the distance measured from the main deck to the inside of the bottom plating, at the midway point of the ship's length.

All British registered dimensions are recorded in feet and tenths of a foot.

Displacement.

The displacement of a ship is the amount of water that is displaced by the ship's hull when she is afloat.

If it is expressed in cubic feet it is called the *Volume Displacement*.

If it is expressed in weight of water (i.e. in tons) it is called the *Weight Displacement*, or more usually, the *Displacement Tonnage*.

Since any floating body always displaces its own weight of the liquid in which it is floating, the displacement tonnage of a ship will be always the same, no matter whether floating in salt or fresh water (provided that there is no alteration in the weight of the ship).

The ship's volume displacement in salt water will be different from that in fresh water, because any volume of salt water is heavier than the same volume of fresh water.

Sea water weighs approximately 64 lb. per cubic foot. In other words, one ton of sea water occupies a volume of 35 cubic feet.

Fresh water weighs approximately 62·5 lb. per cubic foot, so that, one ton of fresh water occupies a volume of 35·84 cubic feet.

River water, at its mouth, marks the change from fresh to sea water and its usually accepted weight is 63 lb. per cubic foot, i.e. one ton of river-mouth water occupies a volume of 35·56 cubic feet.

Suppose that the total weight of a ship and its contents is 8000 tons. Then, in fresh water, its volume displacement would be

$$8000 \times 35{\cdot}84 = 286{,}720 \text{ cubic feet.}$$

Its river-mouth volume displacement would be

$$8000 \times 35{\cdot}56 = 284{,}480 \text{ cubic feet.}$$

Its sea-water volume displacement would be

$$8000 \times 35 = 280{,}000 \text{ cubic feet.}$$

Its weight displacement would, of course, be the same in whatever water it were floating, namely 8000 tons.

Water Plane.

When perfectly calm the surface of the water in which a ship is floating is called the water plane.

Water Line.

The water line of a ship at any time is the line drawn around the hull of the ship to coincide with the water plane.

Load Water Line.

The load water line, or Plimsoll line, is the water line of a ship when carrying its maximum load. It is marked thus: —⊖—.

Freeboard.

The freeboard of a ship is the distance in feet from the main deck to the load water line.

Draught.

A ship's draught is the distance in feet from the lowest part of the keel to the water line at any time.

Draught measurements, at intervals of 1 foot, are shown on the outside of the vessel both at the stem and the stern to indicate the draught forward and aft.

Trim.

The trim of a ship is the difference in feet between the draughts forward and aft.

A vessel drawing the same depth of water at stem and stern, i.e. with the same draught forward as aft, has no trim and is said to be floating on an even keel.

A ship drawing 16 ft. forward and 13 ft. aft is said to trim $(16-13) = 3$ ft. by the bows.

If a vessel trims 4 ft. by the stern, then the draught aft is 4 ft. greater than the draught forward.

Tonnage of Ships.

This is actually a measure of the *capacity* of a ship in cubic feet, the capacity being converted into tonnage by assuming that 100 cubic feet of space is equal to 1 ton weight.

Dock, wharf and canal dues are levied according to a ship's tonnage. This tonnage may be *gross tonnage* or *nett* (or *register*) *tonnage*.

Gross Tonnage.

The sum of the capacities (or volumes) in cubic feet of *all* enclosed spaces of a ship, when divided by 100, is called the ship's gross tonnage. Ballast tanks and double bottoms are *not* included in this calculation. So that a vessel of 1000 tons gross has a cubic capacity, in her combined enclosed spaces (excluding ballast tanks and double bottom, if any), of 100,000 cubic feet.

Nett, or Register, Tonnage.

This is the ship's gross tonnage with a deduction made for (*a*) crew's quarters, (*b*) machinery space.

In merchant vessels and cargo ships the average allowance for crew's quarters is $\frac{1}{20}$th, i.e. 0·05 of the gross tonnage, and machinery space is usually about 0·33 of the gross, making a total deduction of about 0·38 of the gross tonnage.

So that the nett, or register, tonnage of a merchant ship is approximately 0·62 of the gross tonnage. As a general rule, the nett, or register, tonnage of warships is about 0·45 of the gross tonnage.

Load Displacement Tonnage.

When a vessel is loaded to her load water, or Plimsoll, line the number of tons of water so displaced is called the load displacement tonnage (or simply, displacement tonnage). See above.

So that it is (*a*) the total weight of the ship and its maximum cargo, in tons, or, which is the same thing, (*b*) the actual weight in tons of the volume of water displaced below the Plimsoll line.

Dead Weight Tonnage.

This represents the actual carrying capacity of a ship and is the number of tons of cargo that a ship carries when her hull is sunk to the Plimsoll line.

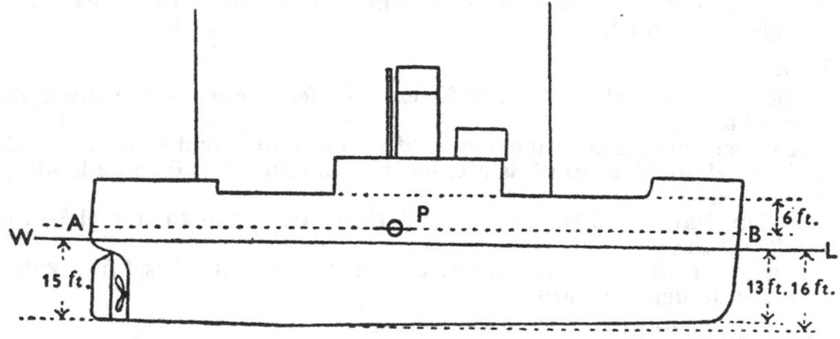

In the above figure *WL* represents the water plane and is consequently the ship's water line.

AB is her load water line and contains the Plimsoll mark *P*.

The draught of the ship is 16 ft. Her forward draught is 13 ft. and the draught aft is 15 ft., so that she trims $(15-13)=2$ ft. by the stern. Her freeboard is 6 ft.

Miscellaneous Problems involving Decimals.

Exercise XXVI.

Assume, for this exercise, that the nett, or register, tonnage of warships is 0·45 of the gross tonnage and that of merchant ships is 0·62 of the gross tonnage.

1. If the gross tonnage of a heavy battleship is 37,240 tons, what is (*a*) her nett register tonnage, (*b*) the total capacity in cubic feet of all enclosed spaces?

2. A light cruiser has a nett register tonnage of 3438 tons. What is (*a*) her gross tonnage, (*b*) the total capacity in cubic feet of all enclosed spaces?

3. The nett register tonnage of a cargo steamer is 2449 tons. What is (*a*) her gross tonnage, (*b*) the total capacity in cubic feet of all enclosed spaces?

4. If the total capacity of all enclosed spaces of a cargo vessel is 725,000 cubic feet, what is her nett register tonnage?

5. The load displacement tonnage of a cruiser is 11,190 tons. Her nett register tonnage is 0·3 of her displacement tonnage. What is (*a*) her nett register tonnage, (*b*) her gross tonnage?

6. When a battleship was refitted her gross tonnage was increased from 27,250 to 31,310 tons. What was the corresponding increase in her nett register tonnage?

7. The nett register tonnage of a heavy cruiser is 4550 tons. If this is 0·28 of her displacement tonnage, what is (*a*) her displacement tonnage, (*b*) her gross tonnage? (Answer to the nearest ton.)

8. From the following details estimate, to the nearest ton, the nett register tonnage of an aircraft carrier: (*a*) Displacement tonnage, 26,500 tons; (*b*) Gross tonnage, 0·54 of displacement tonnage; (*c*) Crew space, 0·05 of gross tonnage; (*d*) Machinery, boilers, bunkers, etc., 0·46 of gross tonnage.

9. The displacement tonnage of a merchant ship is 12,000 tons. If her gross tonnage is 0·45 of her displacement tonnage, what is the capacity in cubic feet of her combined crew and machinery spaces?

10. The displacement tonnage of a destroyer is 3000 tons. If her nett register tonnage is 0·24 of her displacement tonnage, what is the capacity in cubic feet of her combined crew and machinery spaces?

11. A vessel has a nett register tonnage of 5600 tons. Her gross tonnage is 1·4 times her nett register tonnage. If her displacement tonnage is 20,000 tons, (*a*) what decimal is the nett register tonnage of the displacement tonnage, (*b*) what is the total deduction, in tons, for crew and machinery spaces?

If the crew space alone is 47,600 cubic feet, (*c*) what is the deduction in tons for machinery space alone, (*d*) what decimal is this deduction for machinery of the gross tonnage?

12. A merchant ship of 4400 tons nett register has a crew space of 55,000 cubic feet. If a deduction of 0·55 of gross tonnage is allowed for machinery space, what is her gross tonnage?

13. A cargo vessel of 4780 tons nett register has a machinery space of 320,000 cubic feet. If her crew space is 0·05 of her gross tonnage, what is her gross tonnage?

14. If sea water weighs 64 lb. per cubic foot, river-mouth water 63 lb. per cubic foot and fresh water 62·5 lb. per cubic foot, what decimal is (*a*) the weight of 1 cubic foot of fresh water of 1 cubic foot of sea water, (*b*) the weight of 1 cubic foot of river-mouth water of 1 cubic foot of sea water? (Answers to the nearest third decimal place.)

15. A vessel's displacement tonnage is 10,200 tons. If 1 ton of sea water occupies a volume of 35 cubic feet and 1 ton of fresh water a volume of 35·84 cubic feet, what is the ship's volume displacement (*a*) at sea, (*b*) in a fresh-water river port, and (*c*) what decimal is her sea volume displacement of her fresh-water volume displacement? (Answer to three decimal places.)

16. A ship's volume displacement is 179,200 cubic feet in fresh water. What is her displacement in cubic feet when at sea? Does her draught increase or decrease?

17. A ship's volume displacement when at sea is 392,000 cubic feet. (*a*) What is her displacement tonnage? (*b*) What will be the increase in her volume displacement on entering fresh water?

18. If 1 cubic foot of river-mouth water weighs 63 lb., what is the volume, in cubic feet, of 1 ton of river-mouth water? (Answer to the nearest second decimal place.)

19. A ship's volume displacement decreases by 2800 cubic feet on leaving a river mouth and entering the sea. Using the answer obtained in question 18, determine (*a*) her displacement tonnage, (*b*) her sea volume displacement.

20. A cargo ship's volume displacement increases on passing from the sea to a river port (fresh water) by 10,500 cubic feet. If her gross tonnage is 0·54 of her displacement tonnage, what is her nett register tonnage?

THE METRIC SYSTEM

This is used in most countries except the British Empire and U.S.A.

Prefixes. milli = thousandth ($\frac{1}{1000}$).
centi = hundredth ($\frac{1}{100}$).
deci = tenth ($\frac{1}{10}$).
deka = ten times ($10\times$).
hecto = hundred times ($100\times$).
kilo = thousand times ($1000\times$).

Length. Unit: 1 metre = 39·37 inches. (1 in. = 2·54 cm.)
10 millimetres (mm.) = 1 centimetre.
10 centimetres (cm.) = 1 decimetre.
10 decimetres (dm.) = 1 metre.
10 metres (m.) = 1 Dekametre.
10 Dekametres (Dm.) = 1 Hectometre.
10 Hectometres (Hm.) = 1 Kilometre (Km.).

Area. 100 sq. cm. = 1 sq. decimetre.
100 sq. dm. = 1 sq. metre.

Volume. Unit: 1 litre = 1·76 pints.
1000 cubic mm. (c.mm.) = 1 cubic cm. (c.c.).
1000 c.c. = 1 litre.
1000 litres = 1 cubic metre.

Weight. Unit: 1 gram (gm.).
1000 grams = 1 kilogram (kgm.) = 2·204 lb.
1000 kilograms = 1 tonne.

EXERCISE XXVII.

1. (i) What fraction is, (ii) What decimal is:
(*a*) 1 mm. of 1 m. (*b*) 1 m. of 1 Km. (*c*) 1 cm. of 1 m.
(*d*) 3 cm. of 1 dm. (*e*) 4 m. of 1 Km. (*f*) 40 cm. of 1 Km.?

2. Express as kilometres and decimals of a kilometre:
(*a*) 5 Km. 4 m. (*b*) 1 Km. 156 m. (*c*) 56 m.
(*d*) 750 m. (*e*) 2 m. 4 dm. (*f*) 4 m. 46 cm.

3. Convert:
(*a*) 2·462 Km. to cm. (*b*) 2613 c.c. to litres.
(*c*) 4280 gm. to kgm. (*d*) 22 litres to c.c.
(*e*) 76 cm. to metres. (*f*) 466 sq. cm. to sq. metres.
(*g*) 3 tonnes 2 kgm. to gm. (*h*) 66,750 gm. to tonnes.
(*i*) 7 c.mm. to litres. (*j*) 2·4 litres to c.c.

4. Express the following as required:

(a) 7 m. 3 dm. 4 cm. + 9 dm. 3 cm. (answer in m.).
(b) 42 Dm. + 724 mm. (answer in m.).
(c) 125 m. 624 mm. + 71 m. 937 mm. (answer in mm.).
(d) 9·24 Hm. + 16·2 m. + 42,000 mm. (answer in Km.).
(e) 4 kgm. + 321 gm. − 1·6 kgm. (answer in gm.).
(f) 4 cub. metres + 16·5 litres (answer in c.c.).

Conversion of Metric into English Units.

EXAMPLE. If 41 miles = 66 kilometres, convert 182 miles into kilometres (answer to one decimal place). If 41 miles = 66 kilometres, then 1 mile = $\frac{66}{41}$ kilometres.

So that $$182 \text{ miles} = 182 \times \frac{66}{41} = \frac{12{,}012}{41} \text{ Km.} = 292{\cdot}9 \text{ Km.}$$

EXERCISE XXVIII.

1. Convert the following miles into kilometres:
(a) 63. (b) 214. (c) 81·25. (d) 100.

2. Convert the following kilometres into miles:
(a) 100. (b) 571. (c) 32·5. (d) 179.

3. Which coil of rope contains the greater length, and by how many feet?

Coil *A* containing 33·5 fathoms.
Coil *B* containing 61 metres.

4. The length of the earth's equator is 21,600 nautical miles or 40,000 kilometres. Express 1 nautical mile in kilometres (to three decimal places).

5. Convert these litres to gallons (1 litre = 0·22 gal.):
(a) 1000. (b) 600. (c) 450. (d) 2·5 kilolitres.

6. Convert these gallons to litres (1 gal. = 4·546 litres):
(a) 1000. (b) 600. (c) 450. (d) $4\frac{1}{3}$.

7. A kilolitre tank is full but springs a leak. If the loss is a steady 100 c.c. per sec., after how many minutes will the tank be empty?

8. If 350 litres of oil are pumped into a tank in mistake for 350 gal., how many gallons short is the measure?

9. Which is the more efficient pump and by how many gallons per min.?

Pump *A* delivers 360 gal. in 5 min.
Pump *B* delivers 978 litres in 3 min.

10. A man weighs 12 st. 6 lb. What is his weight in kilograms? (Answer to first decimal place.)

11. If petrol weighs 7·2 lb. per gal., find the weight of 1 kilolitre in kilograms.

12. A British standard wire rope of circumference $\frac{3}{4}$ in. has a breaking load of 1·7 tons. Express this fact in metric units, given that 1 in. = 2·54 cm. and 1 metric tonne = 2204·6 lb.

13. The Tay Bridge in Scotland measures 3136 metres between the main abutments, and the Forth Bridge measures 8298·4 ft. Which bridge is the longer and by how many feet?

14. The greatest measured depth in the sea is off Mindanao, in the Pacific Ocean, where a depth of 9783·2 metres was recorded. What is this sounding to the nearest fathom?

15. The greatest measured depth in the Atlantic Ocean is 31,366 feet (Porto Rico Trench). Express this sounding in metres to the first decimal place.

16. (*a*) The height of Mount Everest above mean sea level is 29,141 feet. What is its height in metres? (Answer to first decimal place.)

(*b*) If Mount Everest could be sunk off Mindanao, what depth of water, in fathoms, would there be above its summit? (Answer to nearest fathom.)

17. If the mean depth of the earth's ocean bed is 2000 fathoms below mean sea level, and the mean height of the earth's land surfaces is 701·22 metres above mean sea level, what is the total rise in feet from mean ocean bed level to mean land surface level?

18.

	Displacement	Length	Breadth	Depth
"Majestic"	56,621 tons	915 ft.	100 ft.	58 ft.
"Ile de France"	44,218 tonnes	230·8 m.	27·74 m.	18·6 m.

From the above comparisons find how much the "Majestic" exceeds the "Ile de France" in displacement tons, length and breadth in feet, and how much less she is in depth. 1 tonne = 0·9842 ton and 1 metre = 3·28 ft. (Answer to the nearest unit.)

19. The draught marks on the stem and stern of continental vessels are often shown in both feet and metres (or $\frac{1}{3}$rds of a metre). If the stem of a ship shows a draught of $4\frac{2}{3}$ metres, what is her forward draught to the nearest inch?

20. If a ship's draught is 5·25 metres forward and 18 ft. $10\frac{1}{2}$ in. aft., what is her trim to the nearest inch?

21. A vessel of draught $6\frac{1}{3}$ metres is floating at a place where the depth of water is charted as $5\frac{1}{2}$ fathoms at low-water spring tides (L.W.S.T.). The height of water above charted depth is 12 ft. 7 in. What depth of water, to the nearest inch, is there below the ship's keel?

22. A continental ship of weight displacement 5640 tonnes enters a British port to coal and takes in 360 tons of coal. What is her new weight displacement to the nearest tonne?

23. A British ship of weight displacement 9400 tons takes in 650 tonnes of merchandise at a French port. What, to the nearest ton, is her new displacement tonnage?

24. A British ship of weight displacement 10,520 tons unloads 725 tons of cargo at a French port and takes in 850 tonnes of new cargo. What is her new displacement tonnage to the nearest ton?

25. A bomber sets out to bomb a battleship in dock. The battleship is 205 metres long. Assuming the pilot to make his run up accurately along the fore and aft line of the ship, how many seconds has he in which to drop his bombs to score direct hits, if his speed is 240 miles per hour? (1 metre = 3·28 ft. Answer to nearest tenth of a second.)

FOREIGN WEIGHTS, MEASURES AND CURRENCY

Foreign Weights and Measures.

The accompanying short table of foreign weights and measures, with their corresponding British equivalent, may be referred to for the solution of the following problems.

Country	Unit of weight	Unit of length
United States	1 long ton = 2240 lb. 1 short ton = 2000 lb.	1 mile = 1760 yd.
British India	1 maund = 82·29 lb.	1 koss = 2000 yd.
China	1 picul = 133·3 lb.	1 li = 2115 ft. 1 ch'ih = 1·175 ft.
Egypt	1 kantar = 99·05 lb.	1 dira baladi = 29·83 in.
Turkey	1 kantar = 124·36 lb.	1 arshin = 26·96 in.
Dutch East Indies	1 picul = 136 lb.	—
Russia	1 pood = 36·113 lb. 100 poods = 1·612 tons	1 verst = 0·663 mile

Exercise XXIX.

All answers to be given to the nearest unit.

1. A vessel enters the port of Odessa, Russia, and takes in a cargo of 44,800 poods of wheat in bulk. (*a*) What is the additional displacement tonnage? (*b*) How much stowage space does the cargo take at the rate of 52 cu. ft. per ton?

2. Rice in bags stows at approximately 55 cu. ft. per ton. At Rangoon a grain boat loads 7250 bags of rice each containing 2 India maunds' weight. (*a*) What stowage space will this cargo occupy? (*b*) What is the additional sea-water volume displacement of the ship?

3. A U.S. coaster carries cottonseed from New Orleans. It is stowed in bags each weighing 200 lb. and requires a stowage space of 88 cu. ft. per long ton. If she loads 160 short tons of cottonseed, what stowage space will be needed and how many bags will she take in?

4. A cargo boat loads 3200 baskets of solid Java molasses which stows at 50 cu. ft. per ton. If each basket weighs 1 picul, what stowage space is needed and what is the weight of the cargo in tons?

5. A vessel at Port Said loads 2000 bags of lentils each weighing 1 kantar. What is the ship's additional deadweight tonnage and what stowage space is needed at 54 cu. ft. per ton?

6. A ship at Shanghai takes in a cargo of 1500 bales of silk each weighing 1 picul. What is the ship's extra sea-water volume displacement in cubic feet?

7. A Russian cruiser is steaming at the rate of 40 versts per hour. What is her equivalent speed in knots? (1 N.M. = 6080 ft.)

8. If a Chinese rickshaw boy is paid at the rate of 3*d.* for a journey of 1 li, what would be his fare for a journey of 3¼ miles?

9. In a Turkish bazaar a strip of carpet is sold at the rate of 8*s.* 6*d.* per arshin. What would be its cost per yard?

10. If a length of silk at Shanghai costs the equivalent of 1*s.* 2*d.* per ch'ih, what is its value per yard?

Dominion and Foreign Currency.

Different countries in the world have adopted different coinage systems or currency with different money values and each country has its own particular coinage, or legal tender as it is called, for use within its own borders.

The coinage of each country has what is known as a par value, or standard value, compared with the coinages of all other countries, but by reasons of international finance, national credit and stability, and other considerations, these values may fluctuate. Each country therefore issues from time to time (usually daily) a table of rates of exchange, from which can be seen the values of all foreign monies compared with their own.

It is usually necessary when visiting a foreign country to change your money into the currency of that country, and it is advisable for you to be able to make the calculation involved so as to be satisfied that you have obtained the correct amount and also to enable you to compare prices and values of articles for sale.

There is generally a small charge for changing money, but this is ignored in the following exercise.

Exercise XXX.

Give answers to the nearest halfpenny in English currency.

1. How much should you get by changing £4. 10*s.* 0*d.* into Canadian dollars when the rate of exchange is 4.86 Canadian dollars to the English pound?

2. A man had 18.40 U.S. dollars, spent 7.80 dollars and changed the rest into English money at the rate of 4.84 dollars = £1. What did he receive?

3. In India the standard coin is the rupee. This is divided into annas, so that 1 rupee = 16 annas. If the rate of exchange is $13\frac{1}{3}$ rupees = £1, (*a*) what should you obtain by changing £3. 10*s.* 0*d.* into Indian currency (to nearest anna) and (*b*) what is the value of 1 rupee in English money?

4. At Colombo I had 36 rupees and made purchases to the value of 11 rupees 8 annas. The remainder I exchanged for English currency at the rate of $13\frac{1}{3}$ rupees = £1. What should I have received?

5. The rate of exchange on a certain day in Singapore was 8.57 dollars to the £1. (*a*) What was the value of the Singapore dollar in English money? (*b*) What should a man have received for changing 37.65 dollars into English currency?

6. During a day ashore at Marseilles, with the rate of exchange 124 francs = £1, I spent 270 francs. What was the English equivalent of my day's expenses?

7. If £1 was worth 11.08 Hongkong dollars, what would you have received in English money by changing 24.55 Hongkong dollars?

8. On a day when the rate of exchange at Cairo was $97\frac{1}{2}$ piastres = £1, what should you receive in exchange for £8. 8*s.* 0*d.*?

9. If, when the rate of exchange was $97\frac{1}{2}$ piastres = £1, you bought an article in Port Said for 546 piastres, what was its equivalent price in English money?

10. When the rate of exchange at Barcelona is 31·80 pesetas = £1, (*a*) what would be the equivalent English price of an article costing 63·50 pesetas and (*b*) how much would you receive in exchange for £2. 10*s.* 0*d.*?

11. What should you receive in Russian roubles by changing £5. 4*s.* 0*d.* at Murmansk when the rate of exchange is £1 = 9·45 roubles?

12. What, at the above rate of exchange, is the equivalent English cost of an article priced at 16·50 roubles?

13. A sailor goes ashore at Marseilles and changes £3. 10*s.* 0*d.* into French money at 270 francs to the £1. He spends 285 francs while ashore and returns to his ship with the balance of French money in his pocket. His next port of call is Port Said, where the rate of exchange is £1 = $97\frac{1}{2}$ piastres. What should he receive in piastres for his French money?

14. A visitor to Shanghai bought a silk shawl for 15·25 taels, with the rate of exchange 8 taels = £1. He posted this home from Hongkong as a present and the postage and insurance cost 1*s.* 10½*d.* What change had he out of £2 by making this gift?

15. If the Canadian dollar is 4.95 = £1, and the U.S. dollar 4.84 = £1, what should I get for changing 18.00 Canadian dollars into American dollars?

NAUTICAL MEASURES

The Nautical Table.

The following table of measurement enters into nearly every phase of a seaman's activities and should be committed to memory.

1 nautical mile = 10 cables
= 1000 fathoms
= 2000 yards
1 cable = 8 shackles

Anchoring and Mooring.

The chain which holds the anchor of a vessel is supplied in lengths of one shackle and its size is measured by the diameter of the steel of which the links are made. For large vessels, such as cruisers and battleships, this is usually $2\frac{7}{8}$ in. It is usual for the links of such massive chains to be "studded", as shown in the diagram. This gives additional strength to the chain and prevents it from kinking when stowed.

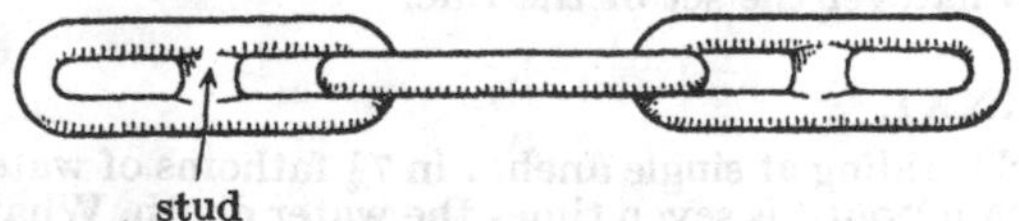

These chains are stowed in a chain locker in the foc'sle near the anchor bed where the anchor is carried suspended from the chain in the bows. The shank of the anchor passes through the hawse-pipe, as shown.

The weight, working load, breaking strain and stowage space of a ship's anchor chain are found from the following simple formulae:

Weight in tons of 100 fathoms of chain of diameter D inches = $2{\cdot}4 \times D \times D$ tons.

Working load of the chain in tons = $18 \times D \times D$ tons.

Breaking strain of the chain in tons = $27 \times D \times D$ tons..

Stowage space for 100 fathoms = $35 \times D \times D$ cubic feet.

Starboard bow of a Light Cruiser

Thus 4 shackles ($\frac{1}{2}$ cable) of 2 in. chain would weigh

$$\frac{1}{2} \times 2{\cdot}4 \times 2 \times 2 = 4{\cdot}8 \text{ tons.}$$

Its working load would be $18 \times 2 \times 2 = 72$ tons.
Its breaking strain would be $27 \times 2 \times 2 = 108$ tons.
Space required for stowage would be $\frac{1}{2} \times 35 \times 2 \times 2 = 70$ cubic feet.

When anchored a vessel may ride either at single anchor, or be moored.

In the first case one anchor only is dropped and a long length of cable, usually 5 to 7 times the depth of water, lies on the sea bottom. In these circumstances, as the tide ebbs and flows, the ship swings with the tide in a circle. The maximum radius of swing of a ship at single anchor is approximately the length of the ship plus the length of chain, centred at the anchor. Obviously, therefore, this kind of anchorage is possible only where there is plenty of sea room.

When a ship is moored, two anchors are dropped, one from the port and one from the starboard hawse-pipe in the bows. While the ship is moving the first anchor is dropped at a certain distance before the required anchorage, and having passed over the proposed spot, the second anchor is dropped at an equal distance beyond it. The chains are then "middled" by being wound in on the capstans so that the vessel lies with her bows at an equal distance from each anchor. For a heavy ship about 6 shackles are used in each chain.

A vessel when moored can swing only in a circle equal to her own length, from a point midway between the anchors. This method thus confines the swing of the ship to a much smaller space.

In a very confined space, such as a busy harbour, a vessel is usually moored both by bow and stern so that there is no appreciable movement of the ship whatever the set of the tide.

Exercise XXXI.

1. A vessel is riding at single anchor in $7\frac{1}{2}$ fathoms of water. The length of anchor chain "out" is seven times the water depth. What is the length of chain in yards?

2. A battleship, at single anchor, in $12\frac{1}{2}$ fathoms of water has six times the water depth of $2\frac{3}{4}$ in. anchor chain "out". What weight is this in tons?

3. A vessel, at single anchor, in $6\frac{1}{2}$ fathoms of water has six times the water depth of $2\frac{1}{2}$ in. anchor chain "out". If the anchor weighs 95 cwt., what is the total weight in tons of the anchor and chain?

4. What is the breaking strain of a battleship's $2\frac{7}{8}$ in. anchor chain? (Answer to the nearest ton.)

5. A vessel 630 ft. long is riding at single anchor in $7\frac{1}{2}$ fathoms of water. The length of anchor chain out is seven times the water depth. What is the approximate maximum radius of swing of the ship in feet?

6. A vessel 825 ft. long is riding at single anchor in $12\frac{1}{2}$ fathoms of water. The length of anchor chain out is five times the water depth. What is the approximate maximum radius of swing of the ship in nautical miles?

7. A ship 480 ft. long is moored by the bow. Both anchors have six shackles of $2\frac{1}{2}$ in. chain out and each anchor weighs 115 cwt. (*a*) What is the total weight of anchors and chains out? (*b*) If the fore and aft line of the ship is in line with the anchors, how far beneath the keel, from the stern, will the nearest anchor lie? (Answer in fathoms, neglecting water depth.)

8. A vessel 525 ft. long, riding at single anchor, has 5 shackles of chain "out". A chart of the anchorage is scaled at 6 in. = 1 N.M. (2000 yd.). With the position of the anchor as centre, what radius in inches should be used to mark on the chart the maximum circle of swing of the vessel's stern?

Soundings.

The taking of a sounding at any place is the determination of the depth of water.

The examination of any marine chart will reveal that waters near the coast have been carefully surveyed with regard to depth of water and also as to the nature of the sea bottom. The numbers recorded indicate fathoms and letters *m*, *s*, *sh*, etc. denote that the sea bottom at those places was mud, sand and shell respectively.

The ebb and flow of the tides, particularly in coastal areas, causes a continual variation in the depth of water at any place. The charted depth is always that taken when the water is at its lowest level, i.e. at low-water spring tides (L.W.S.T.). This level is called the *datum* level. In practice, soundings are taken by one of three methods:

(*a*) the hand lead,
(*b*) the sounding machine,
(*c*) echo sounding.

In the first method, which is used almost exclusively in shallow or coastal waters, the leadsman stands in the "chains"—a platform projecting over the water from the ship's side—and "heaves" the lead forward in the direction of the ship's motion. The lead, itself, is shaped as in the accompanying diagram, with a cup-shaped hollow space in the bottom for "arming the lead", which consists of filling the hollow with tallow so that particles of the material forming the sea bed will adhere to the tallow when the lead reaches the bottom. A practised leadsman can tell at once, by examining the tallow, the nature of the sea bed, and this information is often a useful guide as to a ship's position when visibility is bad.

A good leadsman will heave his lead just so far forward of his position that when the lead has reached the bottom the lead-line, by which he holds it, is as nearly as possible vertical, in which position an accurate reading of the depth of water is obtained. The lead-line is marked in fathoms and the particular mark nearest the surface of the water denotes the depth.

In coastal waters, where tides rise and fall often as much as 18 ft., a correction must be made to soundings to obtain the charted depth, for comparison.

The sounding machine, used in deeper waters where the drag of a lead-line, by reason of the time the lead takes to sink, makes accurate reading difficult, consists of a tube containing a lining of silver chromate. According to the depth to which the machine sinks, the sea water is forced into the tube, by pressure, and the distance it has penetrated is indicated by a white precipitate of silver chloride on the sides of the tube. Reference to a table of measurements issued with the machine determines the corresponding depth in fathoms.

Since the distance the sea water penetrates the tube depends upon the pressure at that depth, a correction must be made for any atmospheric pressure other than normal. The atmospheric pressure for which sounding machines are calibrated is 29·50 in. of mercury.

The correction table is as follows:

For Barometer					
For Barometer	29·75 in.	add 1	fathom	in every 40	fathoms.
,,	30·00 in.	,, 1	,,	30	,,
,,	30·50 in.	,, 1	,,	20	,,
,,	31·00 in.	,, 1	,,	15	,,

EXAMPLE. What is the correct reading of a deep-sea sounding machine showing 2500 fathoms with the barometer at 30·25 in.?

From the above table the correction, to be added, is 1 fathom in every 25 fathoms, i.e.

$\frac{2500}{25}$ fathoms = 100 fathoms to be added.

Thus, the correct depth is 2600 fathoms.

The third method, echo sounding, is the most modern and most accurate of the three, but it necessitates the installation of special apparatus in the ship. This apparatus consists of a sound transmitter, a hydrophone receiver and a time recording machine. The principle involved is, briefly, this. Sound travels at a known speed in sea water (the rate depending upon the "saltness" and temperature of the water) and is reflected back from the sea bottom as an echo.

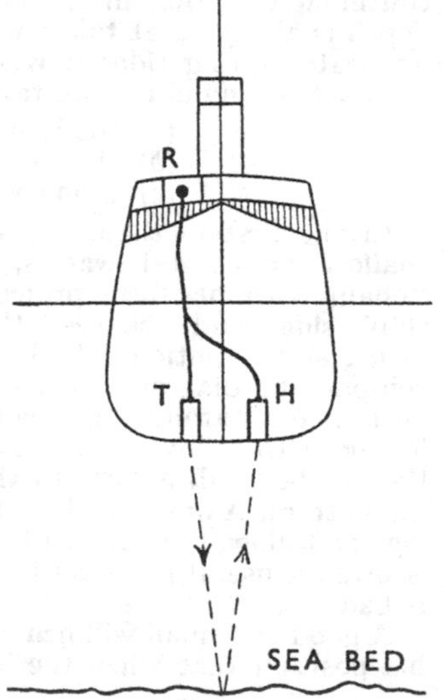

In the above diagram, T is the sound transmitter and H the hydrophone receiver. These are installed in copper cylinders fitted to the ship's bottom plates in a forward position and spaced a small distance each side of the ship's fore and aft line. Heavily insulated electric cables are carried to the recording machine R in the chart room, from which position the whole of the apparatus is controlled. The time taken is evidently that length of time necessary for the sound to travel *twice* the depth of water. The recording machine automatically works out the answer and indicates on a dial the actual depth of the sea bed. One great advantage of this method is that a continuous record can be taken in waters where the bottom is shelving or dangerous.

Exercise XXXII.

1. A sounding of 21 ft. 6 in. is taken while a ship is passing over a bar outside a harbour. The tide is 9 ft. 6 in. above L.W.S.T. What would be the charted depth of water, in fathoms, at the bar?

2. On a ship of draught 16 ft. a sounding of 8 fathoms is taken above a sandbank. If the tide is 14 ft. above datum, what would be the clearance in feet below the ship's keel at L.W.S.T.?

3. A certain rock is charted thus: DRIES 3 ft. This means that when the tide is at datum there are 3 ft. of rock above water. A boat drawing 4 ft. 6 in. passes over this rock when the tide is 10 ft. 6 in. above datum. What is the clearance beneath the boat's keel?

4. A sounding machine recorded a depth of 3800 fathoms with the barometer reading 30·6 in. What was the true depth in fathoms?

5. A recorded depth by sounding machine was 4200 fathoms. If the barometer reading was 30·1 in., what was the true depth in fathoms?

6. If the recorded depth by sounding machine was 3600 fathoms, with a barometer reading of 30·3 in., what was the true depth in nautical miles (2000 yd.)?

7. When using an echo-sounding machine the receiving hydrophone recorded an echo after an interval of $\frac{19}{40}$ sec. If the speed of sound in sea water at that place was 4720 ft. per sec., what was the depth of water to the nearest fathom?

8. If the time interval between transmission and receipt of a sound signal on an echo-sounding machine was 4·32 sec. and the speed of sound in sea water at that place 4650 ft. per sec., what was the depth of water in nautical miles (2000 yd.)? (Answer to three decimal places.)

9. At a certain place the speed of sound in sea water was 4650 ft. per sec. If the depth of water was 620 fathoms, after how long would the receiver of an echo-sounding machine record the echo of a sound signal?

10. In three successive soundings at distances of 2 cables, made by an echo-sounding machine, the time intervals between the signal and the receipt of echo were 1·42 sec., 1·20 sec. and 0·98 sec. respectively. If the speed of sound in the sea water at that place was 4680 ft. per sec., what was the shelving rate of the sea bed in fathoms per nautical mile (2000 yd.)?

The Log-ship and Patent Log.

These instruments, in conjunction with a timing device, indicate the speed at which a ship is moving through the water. They will also show the rate of flow of tide past a stationary vessel at anchor.

The old-fashioned log-ship consisted of a wooden quadrant fitted with an arc of lead to enable it to float upright in the water, as shown.

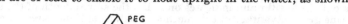

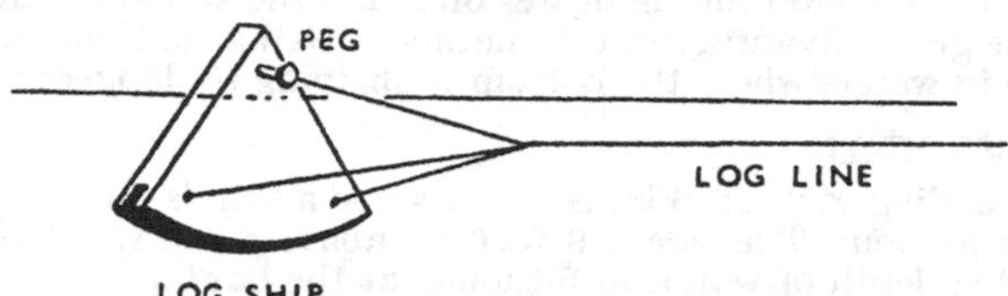

The log-ship was attached to the log line by three short lengths of line, the uppermost of which was fixed to the log-ship by a peg which could be removed by jerking the log line. This enabled the log-ship to be pulled back to the ship, after use, without effort or damage.

The log-ship was cast overboard from the stern of the ship and left floating upright while the log line was "paid out".

This log line was knotted at regular intervals of 47 ft. 3 in. and the number of knots (which for the sake of clarity we will call "marks") that passed over the ship's rail in a given time was noted. The timing device was a "sand glass", similar in construction to an egg-timer, but the time taken for it to "run out" was 28 sec.

The number of marks of the log line that passed over the stern in the 28 sec. was the speed of the ship in knots.

A simple calculation will illustrate that this is so.

If 1 mark in the log line passed over the rail in 28 sec., then the speed of the ship is

$$47 \text{ ft. } 3 \text{ in. in } 28 \text{ sec., i.e. } \frac{47\frac{1}{4}}{28} \text{ ft. in } 1 \text{ sec.}$$

which is

$$\frac{\overset{27}{\cancel{189}}}{\underset{7}{4 \times \cancel{28}}} \times \frac{\overset{15}{\cancel{60}}}{1} \times \frac{\overset{15}{\cancel{60}}}{1} \text{ feet in 1 hour} = 6075 \text{ feet per hour.}$$

This, within the range of practical measurement, is 1 knot.

So that 2 marks on the log line passing over in 28 sec. would represent 2 knots as the ship's speed, and so on.

Usually about 100 ft. of spare line was allowed, before counting the marks and starting the sand glass, to allow for error that might arise from eddy in the ship's wake.

The patent log is an automatic recorder, the dial of which works on somewhat the same principle as a speedometer on a motor car. The principle on which the working of the patent log is based is that of the propeller. When a ship's propeller revolves it tends to move the ship forward by a distance equal to the pitch of the propeller's blades.

Similarly, if a propeller, free to turn, is drawn through the water a distance equal to the pitch of its blades, it tends to revolve once.

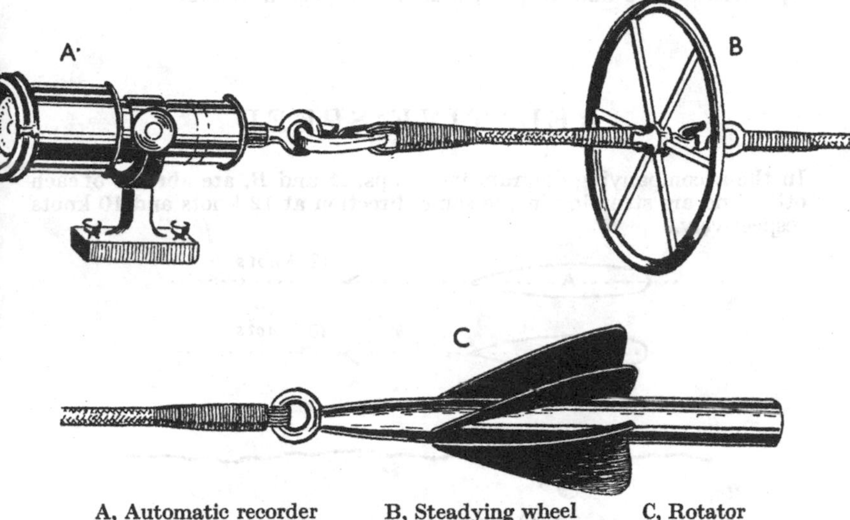

A, Automatic recorder B, Steadying wheel C, Rotator

The accuracy of a patent log depends largely upon the length of line connecting the propeller (or rotator as it is called) to the recorder fixed to the ship's stern rail, and the correct length to suit a certain ship must be ascertained by experiment. The number of revolutions per minute of a ship's engines also enable the speed of the vessel to be calculated with reasonable accuracy, if the ship is not foul, since the principle involved is the same as that used in calculating speed from the patent log.

EXERCISE XXXIII.

1. A log-ship is being used with a sand glass that runs out in 14 sec. What distance apart should the marks be spaced on the log line for the number passing over the stern to record a speed in knots? (Assume 1 N.M. = 6075 ft.)

2. From a stationary vessel a log-ship was used to estimate the speed of the tide. The sand glass used ran out in 36 sec. What distance in fathoms should there be between the marks on the log line so that the number passing over in 36 sec. should record the rate of the tide in nautical miles (2000 yd.) per hour?

3. The log line of a log-ship is knotted at distances of 12½ fathoms. For what length of time should the log line be allowed to run out so that the number of marks passed over records the speed of the vessel in nautical miles (2000 yd.) per hour?

4. A squadron of five ships, each 650 ft. long, is steaming in line ahead with $2\frac{1}{4}$ cables' distance between the bows of successive ships. If the squadron takes 5 min. to pass, what is its speed in knots?

RELATIVE SPEED

In the accompanying diagram two ships, *A* and *B*, are abreast of each other and are steaming in the same direction at 12 knots and 10 knots respectively.

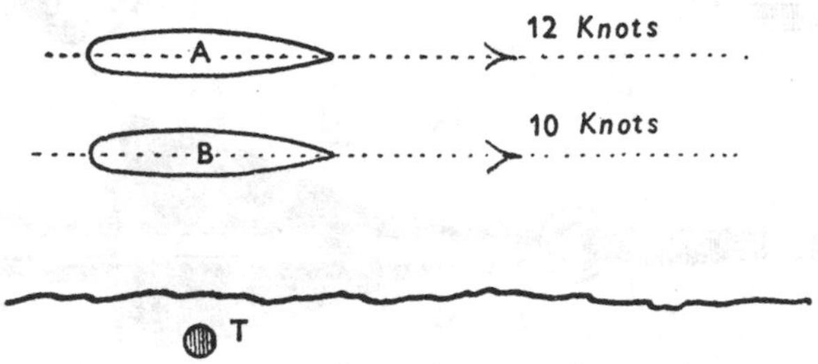

An observer on board ship *A*, looking to starboard, sees ship *B* abreast of his own ship, and, on the shore immediately beyond, he sees the tower *T*.

One hour later, if the two ships continue at the same speed, he sees the tower 12 nautical miles astern and ship *B* 2 nautical miles astern.

His speed, therefore, *relative* to the tower, i.e. a stationary object, is 12 knots, but his speed *relative* to *B* is only 2 knots.

Similarly, the speed of an observer on ship *B* relative to ship *A* is 2 knots *backwards*. Consequently when two vessels are steaming in the same direction their relative speeds, one to the other, are the difference between the two speeds.

If the two ships, *A* and *B*, had been both steaming against a 3 knot tide, then *A*'s speed relative to the land would have been 9 knots, and *B*'s speed 7 knots. The speed of *A* relative to *B* would still be 2 knots forwards, and that of *B* relative to *A* would still be 2 knots backwards.

Had a tide of 3 knots been flowing with them, then *A*'s speed relative to land objects would have been 15 knots and *B*'s speed 13 knots. The relative speed of *A* to *B* and *B* to *A* would still have not changed.

When two vessels are steaming towards one another their relative speed, one to the other, is the sum of their individual speeds.

EXAMPLE 1. Two vessels, steering parallel courses in the same direction, are 20 N.M. apart. The speed of the leading vessel is 14 knots and that of the other is 18 knots. After how long will the second vessel overtake the first?

The speed of the second vessel relative to the first is

$$(18-14)=4 \text{ knots gaining.}$$

So that the time taken to steam the 20 N.M. that they are apart is $\frac{20}{4}=5$ hours.

EXAMPLE 2. Two destroyers, *A* and *B*, are ordered to make contact in the shortest time from two places 128 N.M. apart, using 34 knots and 30 knots respectively. After how long will they meet and how far will the rendezvous be from *B*'s starting point?

The two destroyers are approaching so their relative speed, one to the other, is $34+30=64$ knots.

The time taken for them to meet will be

$$\frac{128}{64}=2 \text{ hours.}$$

The distance *B* will have travelled in 2 hours is 2×30 N.M. $=60$ N.M.

The rendezvous is therefore 60 N.M. from *B*'s starting point.

Gun Range and Range Rate.

When two warships are approaching each other, or when one is overtaking the other, i.e. when the distance between them is shortening, the gun range is said to be "closing".

When the distance between them is increasing, the gun range is said to be "opening".

The rate at which the gun range is opening or closing is known as the "range rate", and is expressed as yards per minute.

For example, if a battleship steaming at $31\frac{1}{2}$ knots is overtaking a cruiser steaming at 30 knots, then the battleship's speed relative to the cruiser is $1\frac{1}{2}$ knots gaining, i.e.

$$3000 \text{ yards in 1 hour} = \frac{3000}{60} \text{ yards in 1 minute.}$$

The range rate, therefore, is 50 yards per minute closing.

EXERCISE XXXIV.

Assume 1 N.M. $=2000$ yd. unless told otherwise.

1. A ship 456 ft. long is steaming at 10 knots. How many seconds will she take to travel her own length? (1 N.M. $=6080$ ft.)

2. Two vessels 18 N.M. apart are steaming in line in the same direction. The leading vessel has a speed of 23 knots and the second vessel a speed of $27\frac{1}{2}$ knots. After how long will the second ship overtake the first and how far will each have travelled in the meantime?

3. A battleship is pursuing a cruiser which is 20 N.M. ahead. The speed of the battleship is $30\frac{1}{2}$ knots and the speed of the cruiser is 28 knots. How long will it take the battleship to get within range if the extreme range of her guns is 22,000 yd.?

4. Two vessels, *A* and *B*, are steaming in the same direction. *A* is 530 ft. long and is steaming at 15 knots. *B* is 382 ft. long and is steaming at 11 knots. If the bow of *A* is level with the stern of *B*, how long will it take *A* to drawn clear of *B*? (1 N.M. $=6080$ ft.)

5. Two ships, A and B, are steaming in opposite directions with their bows abreast. A is 831 ft. long and steaming at 12 knots, while B is 537 ft. long and steaming at 13 knots. How many seconds will they take to pass one another completely? (1 N.M. = 6080 ft.)

6. An armed merchant ship is steaming at 10 knots. At 10 30 hr. a light cruiser, in pursuit at 18 knots, is $12\frac{1}{2}$ N.M. astern. At what time can she open effective fire at 5000 yd. range?

7. A battleship, thought to be an enemy, is reported 170 N.M. away steaming directly towards a seaplane base at 28 knots. If a seaplane sets out at 180 knots and can identify the ship from a distance of 14 N.M., how soon after setting out will the seaplane be able to wireless back her decision?

8. Two seaports, A and B, are 240 N.M. apart. At 10 00 hr. a destroyer leaves A to go to B at 15 knots. At noon a second destroyer is ordered to leave B and join company with the first destroyer, using 20 knots. At what time and how far from A will they meet?

9. At 11 00 hr. a destroyer, steaming at 20 knots, is 20 N.M. astern of her squadron. If she rejoins at 16 00 hr., at what speed is the squadron steaming?

10. A battleship steaming at 16 knots is within 16,000 yd. range of a cruiser. The range rate is 100 yd. per min. opening. If the extreme range of the battleship's guns is 10 N.M., for how long will the cruiser be within range and what is the speed of the cruiser in knots?

11. A number of ships in convoy form a line 5 N.M. in length, steaming at a uniform 10 knots. How long will an escorting destroyer, abreast of the rear ship of the convoy, take to gain a position abreast of the leading ship, if she uses 22 knots?

12. Two enemy battleships, A and B, are coming into action and steaming towards each other. They are 12 N.M. apart at noon. A is steaming at 16 knots and decides to open fire at 16,000 yd. on B which is steaming at 14 knots. What is the range rate and at what time will A fire her first salvo?

13. Three destroyers, A, B and C, are undergoing speed trials. A beats B by 5 cables in 10 N.M. and beats C by $7\frac{1}{2}$ cables over the same distance. By how much would B beat C over a distance of 38 N.M.?

14. Two destroyers leave the same port steering the same course. The first destroyer leaves 20 min. in front of the other, steaming at 26 knots. The second destroyer uses 24 knots. How far apart are they one hour after the departure of the first destroyer?

15. Two destroyers leave the same port steering the same course. The first destroyer leaves 12 min. in front of the other, steaming at 20 knots. The second destroyer uses 23 knots. How far apart are they one hour after the departure of the second destroyer?

16. A battleship $5\frac{1}{2}$ N.M. off shore, steaming at 24 knots, fires a heavy gun. By how far, in cables, has the battleship changed position when the sound of the discharge is heard on shore? (Sound travels at 1100 ft. per sec.)

17. A river patrol boat is capable of a speed of 12 knots in still water and can carry enough fuel for 10 hr. steaming at that speed. At 13 30 hr. she sets off downstream with a 3 knot current. At what time must course be altered to return so as to be able to reach her base?

18. A motor boat carries enough fuel for 6 hr. cruising at 12 knots in still water. How far upstream may she travel at this speed against a 3 knot current so as to be able to return under her own power?

19. A battleship is pursuing a cruiser, and their respective speeds are 18 knots and 16 knots. How far will each vessel steam while the gun range is closing from 16,000 to 14,000 yd.?

20. A ship is in position *B*, which is 69 N.M. downstream from another ship in position *A*. The tide is flowing at 3 knots. At 10 00 hr. the vessel at *A* sets out at 12 knots to meet the vessel from *B*. At 10 12 hr. the vessel at *B* steams out at 10 knots to meet the vessel from *A*. At what time will they meet and how far from *A*?

AVERAGES

Suppose that a ship left Cape Town bound for Adelaide and on four successive days her noon to noon runs were 391, 416, 435 and 426 nautical miles.

Her total mileage travelled would then be 1668 N.M. and her *average* daily run, from noon to noon, would be found by dividing this distance by the number of days, i.e. $\frac{1668}{4} = 417$ N.M.

(The word *mean* is sometimes used instead of average.)

In the same way her *average* speed for the *whole* four days is

$$\frac{1668}{96 \text{ hours}} = 17{\cdot}375 \text{ knots.}$$

It would be incorrect to say that her average speed for *any* of the four days was 17·375 knots. When we say that the average of 16 knots, 12 knots and 17 knots is $\frac{16+12+17}{3} = 15$ knots, we assume the ship spent an equal time at each speed.

When a ship travels at different speeds for different times the average speed can be obtained only from the total distance run and the total time taken.

Example. What was the average speed of a ship that steamed 12 knots for 3 hours, 17 knots for 4 hours and 16 knots for 5 hours?

Distance travelled in 3 hr. at 12 knots = 36 n.m.
,, ,, 4 hr. ,, 17 knots = 68 n.m.
,, ,, 5 hr. ,, 16 knots = 80 n.m.

Therefore

Distance travelled in $(3+4+5)$ hr. $=(36+68+80)$ n.m.

i.e. 184 n.m. in 12 hr.

The average speed, therefore, is $\frac{184}{12}=15\frac{1}{3}$ knots.

Exercise XXXV.

1. Find the average daily atmospheric pressure in millibars from these daily records for one week: 1001·6; 1006·2; 1008·7; 1017·6; 1004·8; 1002·5; 999·9.

2. Having steamed for $4\frac{1}{2}$ hr. at 18 knots a ship reduced speed to 16 knots and covered 96 n.m. at that speed. Speed was then again reduced to 12 knots for $1\frac{1}{2}$ hr. What was her average speed for the journey?

3. The average barometer reading for seven consecutive days was 29·83 in. of mercury. If the average for the first four days was 29·91 in., what was the average for the last three days?

4. A ship sails from latitude 42° 30′ N. to latitude 50° 16′ N. What is the mean latitude of these two positions?

5. What is the mean latitude between 18° 30′ N. and 11° 30′ S.?

6. What is the mean time and the mean sextant altitude from the following solar observations?

Time:	11 hr. 31 min. 20 sec.;	11 hr. 32 min. 10 sec.;	11 hr. 32 min. 45 sec.
Altitude:	54° 20′ 32″;	54° 20′ 50″;	54° 21′ 20″.

7. A ship steamed at 12 knots for 2 hr. 20 min., at 16 knots for 3 hr. 15 min., and at 18 knots for 1 hr. 25 min. What was her average speed? (Answer to nearest first decimal place.)

8. In a dockyard the following numbers of 6 in. bolts were used on six consecutive working days: 350; 280; 410; 296; 376; 346. Estimate, to the nearest gross, the requirements for a period containing 78 working days.

9. The numbers of men accommodated daily in a ship's mess, for the seven days of one week, were: 36; 34; 38; 27; 30; 28; 31. If 36 oz. of food is the daily ration for each man, what was the average weight in lb. of food supplied daily?

10. The average weight of a boat's crew of 8 men was originally 11 st. 3 lb. Before the race took place a man weighing 10 st. 7 lb. went sick and his place was taken by a man weighing 11 st. 9 lb. What was the new average weight of the crew?

11. A boat's crew consisted of 8 oarsmen and a coxswain. The average weight of the oarsmen was 11 st. $1\frac{1}{2}$ lb. The average weight of the whole crew including the coxswain was 10 st. 13 lb. What was the weight of the coxswain?

12. The following were the actual heights of high and low water at Devonport on three successive days:

H.W.:	13·5 ft.;	13·3 ft.;	14·1 ft.;	14·0 ft.;	15·0 ft.;	14·8 ft.
L.W.:	3·3 ft.;	2·9 ft.;	2·3 ft.;	1·6 ft.;	0·7 ft.;	0·4 ft.

Find the average difference in height between high and low water for these six tides.

RATIO

A ratio is used to express a relationship between any two or more quantities of the same kind.

For example, if a vessel has a maximum speed of 24 knots, but her most economical cruising speed is 18 knots, then the ratio of her cruising speed to her maximum speed is written thus:

$$18:24.$$

This may also be expressed as a fraction:

$$\tfrac{18}{24} = \tfrac{3}{4}, \text{ in its lowest terms,}$$

or as a decimal $= 0·75$.

Notice that in all the following cases the ratio is the same, namely 3:4.

	Cruising speed	Maximum speed
(*a*)	12	16
(*b*)	15	20
(*c*)	21	28
(*d*)	39	52

Hence we see that the ratio expresses only the relationship between the speeds and is not itself a speed, nor, in fact, any quantity.

When two quantities bear a known ratio to each other and the value of one is known, then the value of the other can be readily obtained.

EXAMPLE 1. In a particular ship the ratio of length to breadth was 17:2. Her actual length was 391 ft. Find her breadth.

Since the ratio Length:Breadth = 17:2,

then the ratio Breadth:Length = 2:17,

i.e. $$\frac{\text{Breadth}}{\text{Length}} = \frac{2}{17}.$$

Therefore $$\text{Breadth} = \frac{2 \times \text{Length}}{17} \text{ ft.} = \frac{2 \times \overset{23}{\cancel{391}}}{\cancel{17}} = 46 \text{ ft.}$$

In any ratio the quantities expressed must be always in the same units.

EXAMPLE 2. The length of the S.S. "Mauretania" is 790 ft. and the length of the liner "Normandie" is 313·7 metres. What is the ratio of the length of the "Mauretania" to the length of the "Normandie"?

The ratio, when both are expressed in feet, is

$$\frac{\text{"Mauretania"}}{\text{"Normandie"}} = \frac{790 \text{ ft.}}{313{\cdot}7 \times 3{\cdot}28 \text{ ft.}} \text{ (since 1 metre} = 3{\cdot}28 \text{ ft.)}$$

$$= 0{\cdot}767 \text{ (by working).}$$

The ratio, when both are expressed in metres, is

$$\frac{\text{"Mauretania"}}{\text{"Normandie"}} = \frac{790 \times 0{\cdot}3048 \text{ metres}}{313{\cdot}7 \text{ metres}} \quad \text{(since 1 ft.} = 0{\cdot}3048 \text{ metre)}$$

$$= 0{\cdot}767 \text{ (by working).}$$

EXERCISE XXXVI.

1. Find the value of the missing quantities in the following table, where the ratios represent Cruising speed : Maximum speed:

	Ratio	Cruising speed	Maximum speed
(*a*)	3:4	—	22 knots
(*b*)	$\frac{11}{16}$	$27\frac{1}{2}$ knots	—
(*c*)	0·75	24 knots	—
(*d*)	0·83	—	35 knots

2. A chart is so drawn that a distance on the ground of $3\frac{1}{4}$ N.M. is represented by a length of $4\frac{1}{3}$ in. on the chart. What is the ratio of chart distance to actual distance expressed in the same units? (1 N.M. = 6080 ft.)

3. The length of the S.S. "Winchester Castle" is 630 ft. and the breadth is 73 ft. 6 in. What is the breadth of a scale model of this ship if its length is 2 ft. 3 in.?

4. The weights of 1 cubic foot of water are as follows: fresh water, $62\frac{1}{2}$ lb.; sea water, 64 lb.; river-mouth water, 63 lb. What is the weight ratio of an equal volume of (*a*) sea water to fresh water, (*b*) river-mouth water to sea water? (Answers to the nearest second decimal place.)

Safe Working Load Ratios for Ropes.

All ropes have a *breaking strain*, which is expressed as the smallest weight that will break the rope.

The safe working load must obviously be considerably less than the breaking strain and these are usually expressed as a ratio.

Thus, the ratio $\frac{\text{Safe working load}}{\text{Breaking strain}} = \frac{1}{6}$ for wire ropes.

The ratio $= \frac{1}{8}$ for dry, untarred ropes.

If the breaking strain for a rope is known, the safe working load may be easily calculated.

Sometimes, when it is not desirable to exert a breaking strain, a rope is subjected to what is called a *testing*, or *proof load*, which depends upon the size of the rope. If the rope withstands this test, the safe working load is then expressed, with this proof load, as a ratio.

The ratio $\frac{\text{Safe working load}}{\text{Proof load}} = \frac{1}{2}$ for dry, untarred rope.

When rope is wet, or tarred, its safe working load is reduced by $\frac{1}{4}$.

When a rope has been spliced, its safe working load is reduced by $\frac{1}{8}$ for each splice.

Exercise XXXVII.

1. Find, to the nearest second decimal place, the safe working load, in tons, of the following wire ropes: (*a*) breaking strain, 2·95 tons; (*b*) breaking strain, 4·45 tons; (*c*) breaking strain, 11·85 tons.

2. What is the safe working load of a dry, untarred rope of breaking strain (*a*) 6·4 cwt., (*b*) 16·4 cwt.?

3. What is the safe working load of a dry, untarred rope whose proof load is 1 ton 16 cwt.? What would be its breaking strain?

4. A dry, untarred rope has a safe working load of 1 ton 8 cwt. If this rope is wetted, under what load would it probably break?

Pulley Ratios.

The lifting of weights by pulleys or tackle is an everyday operation in the working of any ship, so that pulley ratios are very important.

These ratios are usually expressed as $\frac{\text{Effort expended}}{\text{Load lifted}}$, and are known as "effort-load" ratios.

The following are the effort-load ratios for a few of the more common tackles or pulley systems:

(*a*) Double whip.
Ratio 1:2.

(*b*) Gun tackle.
Ratio 1:2.

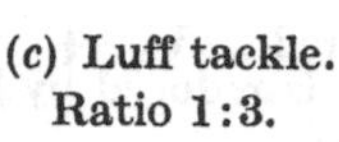

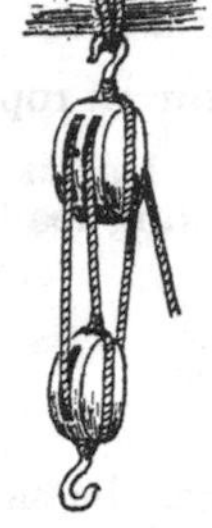

(*c*) Luff tackle.
Ratio 1:3.

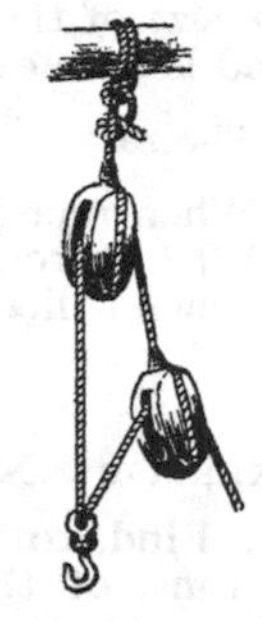

(*d*) Single Spanish Burton.
Ratio 1:3.

(*e*) Double luff.
Ratio 1:4.

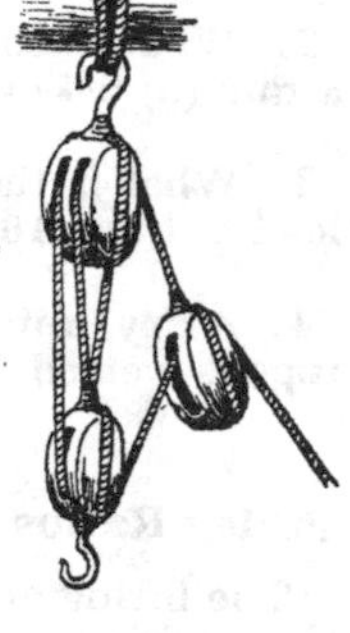

(*f*) Double Spanish Burton.
Ratio 1:5.

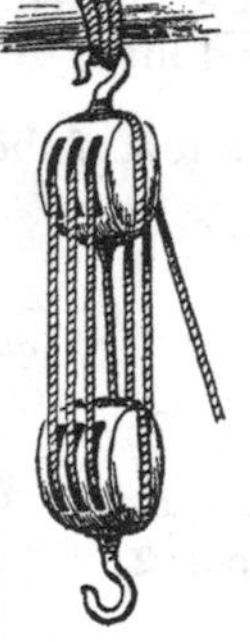

(*g*) Threefold purchase.
Ratio 1:6.

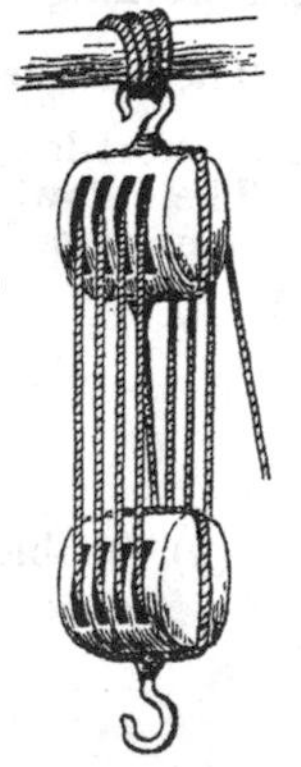

(*h*) Fourfold purchase.
Ratio 1:8.

In practice these ratios are never fully established because of the friction of the rope and sheaves of the blocks. This will be considered later, but, for the following exercise, assume the above ratios to hold good.

Exercise XXXVIII.

1. What load, in cwt., could be lifted with double-whip tackle when exerting a pull of 84 lb.?

2. What effort is required to lift a load of 2 tons 5 cwt. when using single Spanish Burton?

3. A man weighing 12 st. 7 lb. has to lift a load of 12 cwt. by using overhead tackle. Which of the above systems could he use?

4. A dry untarred rope whose normal safe working load is 15 cwt. is being used in overhead threefold purchase tackle by a man able to exert a pull equal to his weight of 15 st. The rope has been spliced in one place. Will the safe working load be exceeded and, if not, what is the margin of safety in lb.?

Division of Quantities in a Given Ratio.

It is sometimes necessary to divide a quantity into a number of parts, which parts must bear a certain ratio to one another.

Example 1. 2400 tons of cargo are to be shared between three ships in the ratio of 2:3:7. What tonnage should each ship receive?

By adding together the numbers in the ratio, $2+3+7=12$, we obtain the total number of "shares".

So that the first ship receives 2 shares, i.e.

$\frac{2}{12}$ of the cargo $=\frac{2}{12}\times 2400$ tons $=400$ tons.

The second ship receives 3 shares, i.e.

$\frac{3}{12}$ of the cargo $=\frac{3}{12}\times 2400$ tons $=600$ tons.

The third ship receives 7 shares, i.e.

$\frac{7}{12}$ of the cargo $=\frac{7}{12}\times 2400$ tons $=1400$ tons.

A further development of the above example is to be found in cases where there are more than one of some of the "participants".

Example 2. 2310 tons of cargo are to be divided between sailing barges, lighters and dumb-barges so that the ratio between sailing barge, lighter and dumb-barge is 10:8:7. If there are 1 sailing barge, 4 lighters and 5 dumb-barges to be loaded, what tonnage of cargo should be assigned to each type of vessel?

In this problem the total number of "shares" is found by multiplying the respective figure in the ratio by the number of the corresponding type of vessel. Thus:

1 sailing barge receives $1\times 10=10$ shares
4 lighters receive $4\times 8=32$,,
5 dumb-barges receive $5\times 7=35$,,

The total number of shares is, therefore, $10+32+35=77$.

The weight of each share is $\frac{2310}{77}=30$ tons.

Therefore the sailing barge receives $10\times 30=300$ tons
a lighter receives $8\times 30=240$,,
a dumb-barge receives $7\times 30=210$,,

Rise and Fall of Tides.

The rise and fall of tides afford an exercise in division in a given ratio.

Although the time between low water and the next succeeding high water at a place may be, say, 6 hr., the rate of increase of depth of water is not uniform. In other words, the water does not rise the same amount for every hour of the six.

The method employed to determine the rate of rise or fall of tide is illustrated in the following diagram.

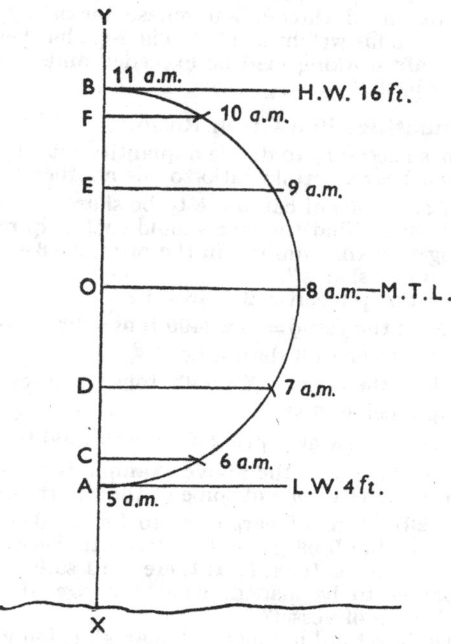

Example 3. Suppose that low water is at 5 a.m. and the depth of water above sea bottom is then 4 ft. Assume also that the next high tide is at 11 a.m., when the depth of water is 16 ft.

The total rise of tide for 6 hr. is, therefore, 12 ft.

Draw a perpendicular line XY and let the point A, in the diagram, represent the depth (4 ft.) at low water at 5 a.m. Let the point B represent the depth (16 ft.) at high water at 11 a.m. The point O, midway between A and B, will then represent the mean tide (or half tide) level.

With centre O and radius OA describe a semicircle. By means of compasses or a protractor divide this semicircle into six equal divisions each representing one hour. These are labelled 5 a.m., 6 a.m., 7 a.m., ... 11 a.m. as shown.

From the hour marks, 6 a.m., 7 a.m., 9 a.m., 10 a.m. draw horizontal lines to meet the line *AB* at *C*, *D*, *E* and *F*.

Then the lengths *AC*, *CD*, *DO*, *OE*, *EF* and *FB* represent the rise of tide for each hour respectively from 5 a.m. to 11 a.m.

By measurement, or by geometrical calculation, these lengths may be shown to be, as nearly as possible, in the ratio 1:3:4:4:3:1.

The sum of these numbers is 16.

Therefore the tide rises $\frac{1}{16}$ of its total in the first hour, from 5 a.m. to 6 a.m.,

$$=\tfrac{1}{16}\times 12=\tfrac{3}{4}\text{ ft.}=9\text{ in.}$$

In the second hour, from 6 a.m. to 7 a.m. the rise is

$$\tfrac{3}{16}\times 12=\tfrac{9}{4}\text{ ft.}=2\text{ ft. }3\text{ in., and so on.}$$

Exercise XXXIX.

1. A length of rope measuring 25 fathoms has to be divided into three lengths in the ratio 3:5:7. What is the length in feet of each piece?

2. A mass of Admiralty gun-metal weighing $37\frac{1}{2}$ cwt. is found, upon analysis, to contain copper, tin and zinc in the ratio 44:5:1. What are the weights of each metal used to make this?

3. What would be the length of each piece of a steel rod 6 ft. 8 in. long if divided into three pieces in the ratio 2:5:9?

4. 1 cwt. of Admiralty bronze containing 45 parts by weight of copper, 10 of nickel and 45 of zinc is blended with 2 cwt. of a nickel brass containing 60 parts by weight of copper, 15 of nickel and 25 of zinc. What is the ratio of nickel to zinc in the new alloy?

5. During refit a certain job was done in a destroyer by three men, of different trades, but each earning the same hourly rate of pay. The first man worked 5 hr., the second 7 hr. and the third 12 hr. The total payment in wages for the job was £2. 14*s*. 0*d*. What did each man receive?

6. Low water at Chatham was at 3 a.m. on 15 May and the depth of water was 8 ft. At the next high tide, at 9 a.m., the depth was 24 ft. What depth of water was there at 7 a.m.?

7. At a certain place where low water was at noon the beach was 2 ft. 6 in. above the water level. If high water was at 6 p.m. and the beach was then covered to a depth of 11 ft. 6 in., what depth of water was there at 2 p.m.?

8. On a certain day when low water was at 2.30 p.m. the level of the water rose 1 ft. 9 in. between 3.30 p.m. and 4.30 p.m. What was the range of the tide and how much did the water level rise between 4.30 p.m. and 5.30 p.m.?

9. A cargo of 3915 tons is to be divided among coasters, sailing barges and lighters so that the ratio between a coaster, sailing barge and lighter is 12:7:5. If there are 2 coasters, 4 sailing barges and 7 lighters to be loaded, what tonnage of cargo will each type of vessel receive?

10. A bonus amounting to £590 was shared between the following ratings so that the ratio between petty officer, leading seaman, able seaman and ordinary seaman was 11:7:5:4. If there were 3 P.O.s, 5 Ldg. seamen, 2 A.B.s and 10 ordinary seamen in the "share out", how much did each receive?

PROPORTION

Suppose we wish to construct a model of a ship whose length is 450 ft. and whose breadth is 50 ft. and we decide to make the model 3 ft. long.

Then the length of the model is $\frac{3}{450}=\frac{1}{150}$ of the actual length of the ship; and for the model to be true to scale the breadth of the model must be $\frac{1}{150}$ of the breadth of the ship:

$$=\frac{50 \text{ ft.}}{150}=\frac{1}{3} \text{ ft.}=4 \text{ in.}$$

All other measurements must be $\frac{1}{150}$ of full size and they are then all in the same ratio and are said to be *in proportion*.

Hence we can say that the ratio of the model's length to the ship's length is equal to the ratio of the model's breadth to the ship's breadth.

This is usually written in this form:

Model's length : Ship's length = Model's breadth : Ship's breadth;

or
$$\frac{\text{Model's length}}{\text{Ship's length}}=\frac{\text{Model's breadth}}{\text{Ship's breadth}}.$$

It is equally true to say:

$$\frac{\text{Model's length}}{\text{Model's breadth}}=\frac{\text{Ship's length}}{\text{Ship's breadth}}.$$

It is obvious, therefore, that if three of these four quantities are known, the fourth can be readily obtained.

Direct Proportion.

Consider the case of distance travelled by a ship at sea. The longer she is under way the farther she travels. If her speed is steady and does not change, then she travels three times as far in 3 hr. as she does in 1 hr. So that for any increase or decrease in time taken there is a corresponding increase or decrease in distance run, and vice versa.

These quantities are in *direct proportion*.

EXAMPLE 1. If a ship travels $10\frac{1}{2}$ N.M. in 24 min., what is her speed in knots?

In other words, how far will she travel in 1 hr. at the same speed?

Unit Method.

In 24 min. the ship travels $10\frac{1}{2}$ N.M.

Therefore in 1 min. ,, ,, $\frac{10\frac{1}{2}}{24}$ N.M.

And in 60 min. ,, ,, $\frac{60\times 10\frac{1}{2}}{24}$ N.M.

$$=\frac{\overset{5}{\cancel{60}}\times 21}{\underset{2}{\cancel{24}}\times 2}=\frac{105}{4}=26\tfrac{1}{4} \text{ N.M.}$$

Her speed, therefore, is $26\frac{1}{4}$ knots.

Fractional Method.

The greater the time taken, the greater the distance travelled. Therefore the new distance will be $\frac{60}{24}$ times the old distance

$$=\tfrac{60}{24}\times 10\tfrac{1}{2}=26\tfrac{1}{4}\text{ N.M.}$$

EXAMPLE 2. A vessel loads 360 tons of fuel oil in $2\frac{1}{2}$ hr. How long will she take to load 540 tons?

Unit Method.

360 tons of oil are loaded in $2\frac{1}{2}$ hr.

Therefore 1 ton " is " $\dfrac{2\frac{1}{2}}{360}$ hr.

And 540 tons " are " $\dfrac{2\frac{1}{2}\times 540}{360}$ hr.

$$=\frac{5\times \overset{3}{\cancel{540}}}{2\times \underset{2}{\cancel{360}}}=\frac{15}{4}\text{ hr.}=3\tfrac{3}{4}\text{ hr.}$$

Fractional Method.

More oil will take longer to load, so that the new time is $\frac{540}{360}$ of the old time:

$$=\tfrac{540}{360}\times 2\tfrac{1}{2}=3\tfrac{3}{4}\text{ hr.}$$

EXERCISE XL.

1. From the following distances travelled in given times find the corresponding speed in knots, to the nearest first decimal place where necessary:

	Time taken	Distance run (N.M.)
(*a*)	$7\frac{1}{2}$ min.	2
(*b*)	$12\frac{1}{4}$ min.	$3\frac{1}{2}$
(*c*)	1 hr. 16 min.	$24\frac{1}{2}$
(*d*)	12 hr. 20 min.	182

2. Distance travelled when estimated from clock time taken and the speed in knots is called "dead reckoning" (D.R.).

Find the D.R. distances from the following times and speeds, correct to nearest first decimal place:

	Time taken	Speed in knots
(*a*)	1 hr. 35 min.	18
(*b*)	3 hr. 21 min.	$12\frac{1}{2}$
(*c*)	9 hr. 6 min.	10·7
(*d*)	27·5 hr.	11·8

3. While burning 128 tons of fuel a vessel covers a distance of 252 N.M. How far, at the same rate of fuel consumption, can she run on her total bunker capacity of 640 tons?

4. From the following speeds and distances travelled determine the time taken, to the nearest min.:

	Speed in knots	Distance in N.M.
(*a*)	12	135
(*b*)	$10\frac{1}{2}$	$181\frac{1}{2}$
(*c*)	15·3	201·8
(*d*)	21·6	324·3

5. A ship uses 72 tons of bunker coal in $13\frac{1}{2}$ hr. What is her daily consumption at this rate and how long can she stay at sea if her bunker capacity is 480 tons?

6. If a ship can steam from one port to another in 36 hr. at 18 knots, (*a*) at what speed must she steam to arrive in 24 hr., (*b*) how long would she take at 24 knots?

7. If 155 French francs are equal in value to £1. 5*s.* 0*d.*, how many francs would be given in exchange for £3. 10*s.* 0*d.*?

8. If 24 rupees are worth 36 shillings, change (*a*) 17 shillings into rupees, (*b*) 42 rupees into shillings.

9. If an 84 lb. pull on a lifting tackle can raise a weight of 756 lb., what weight could be raised by a pull of $\frac{1}{2}$ cwt.?

10. The Board of Trade regulations with respect to pontoon lifeboats ensure that a boat 28 ft. long shall clear 2 tons of water from her decks in 60 sec. and for all other boats the weight of water cleared in 60 sec. must be directly proportional to the length.

How many tons of water must pontoon lifeboats of the following lengths be able to clear from their decks in one minute?

(*a*) 32 ft. (*b*) 24 ft. (*c*) 36 ft.

Inverse Proportion.

In some cases we find that as one quantity increases, so the other decreases.

For instance, the *more* men employed to do a job, the *less* the time taken to do it.

The *greater* the speed of a ship, the *less* time it takes for the journey. Such cases are examples of *inverse proportion.*

EXAMPLE 1. A boat's crew of 16 men have enough food to last for 18 days. How long would the same amount of food last 24 men?

Unit Method.

The food supply lasts 16 men for 18 days.

Therefore the food supply lasts 1 man for 16×18 days.

And ,, ,, 24 men ,, $\frac{16\times 18}{24}$ days $=12$ days.

Fractional Method.

The greater the number of men, the shorter the time the food will last. Therefore the new time is $\frac{16}{24}$ of the old time

$$=\tfrac{16}{24}\times 18=12 \text{ days.}$$

Problems in proportion are sometimes encountered where we have more than three varying factors to consider involving possibly both direct and inverse proportion.

These are best solved by the fractional method in two separate stages.

EXAMPLE 2. A boat contains 12 survivors and carries sufficient food for 6 days if the daily ration per person is 15 oz. If 6 additional survivors are picked up, how long will the provisions last if the daily ration per person is reduced to 12 oz.?

First, we have an increase in the number of survivors, from 12 to 18, which reduces the time the rations will last to $\frac{12}{18}$ of the original time, i.e.

$$\text{The new time}=\tfrac{12}{18}\times 6 \text{ days.}$$

Secondly, this new time will be affected by the altered rations.

Reducing the rations from 15 oz. to 12 oz. per head will increase the final time to $\frac{15}{12}$ of the new time, i.e.

$$\text{The final time}=\tfrac{12}{18}\times\tfrac{6}{1}\times\tfrac{15}{12}=5 \text{ days.}$$

Therefore the time that the provisions will last, under the new conditions, is 5 days.

EXERCISE XLI.

1. A vessel with a complement of 750 men carries food sufficient for 17 days. If 100 more men join the ship before sailing, how long will the same quantity of provisions last?

2. A ship can stay at sea for 12 days if her fuel consumption is 150 tons per day. By how many days will her time at sea be reduced if she burns 180 tons per day?

3. A vessel steaming at 15 knots would arrive at her destination in 18 hr. By how many knots must she increase speed to arrive in $12\frac{1}{2}$ hr.?

4. A ship is under orders to reach a rendezvous 180 N.M. distant in 12 hr. She sets out at 7 a.m. steaming at $17\frac{1}{2}$ knots. By how many knots must her speed be reduced at 9 a.m. so as to arrive on time?

5. If 480 men can coal ship in 6 hr., how many additional men are needed to reduce the coaling time by 2 hr.?

6. A boat's crew of 12 men have sufficient food to last 12 days if each man is allowed 12 oz. per day. At the end of 5 days they find they are still 9 days' sail from land. What must be the ration per man per day for the last 9 days so that the food will last until the end of the voyage?

7. The speed of a ship fully laden is 15 knots with a steady coal consumption of 120 tons per day. If her speed increases by $\frac{1}{2}$ knot for every 80 tons' reduction in her displacement tonnage, what is her speed at the end of the fifth day?

8. A ship can stay 18 days at sea when burning 176 tons of fuel per day. What must be her daily consumption to allow an increase of 4 days on her cruising time?

9. A shipwrecked crew of 10 men in a ship's boat allow themselves $\frac{1}{2}$ lb. of food each per day and estimate their stock will then last 2 weeks. After 8 days adrift they decide for safety to make their remaining rations last another 12 days. What amount should be allotted to each man daily?

10. Two objects represented on a chart scaled at 4 in. = 1 N.M. measure 3·8 in. apart. What distance apart would they measure on a chart of which the scale is 2 N.M. = 1 in.?

11. A boat containing 8 men has a water supply that will last 12 days with a daily ration of $1\frac{1}{2}$ pints per man. If 4 additional men are picked up, what must be the daily ration of water for the supply to last 16 days?

12. If 1750 men can build a breakwater 2250 yd. long in 24 days, how long will it take 560 men working at the same rate to build a breakwater 480 yd. long?

13. A shipbuilding company contracts to build a light cruiser in 300 working days, employing 1600 men. After 120 working days the cruiser is only a quarter completed. How many more men, working at the same rate, must be employed to finish the contract to time?

14. Certain engine room repairs are estimated to take 14 men, each working 8 hr. a day, 19 days to complete. At the end of 10 days an accident occurs and 4 men are killed and all work is suspended for 2 days. How many more men must be put on to finish the work to time if they all now work 9 hr. a day?

PERCENTAGE

A percentage is simply a ratio in which the second term is always 100.

For example, 80 per cent (the sign for "per cent" is %) means a ratio of 80:100, i.e. 80 hundredths.

Any fraction may be expressed as a percentage by multiplying it by 100.

Thus $\frac{4}{5}$, expressed as a percentage, is $\frac{4 \times 100}{5} = 80\ \%$.

And $\frac{7}{16}$, expressed as a percentage, is $\frac{7 \times 100}{16} = 43\frac{3}{4}\ \%$.

Any decimal fraction may be expressed as a percentage by multiplying it by 100, i.e. by moving the decimal point 2 places to the right.

Thus 0·25 is 25 %, and 3·21 is 321 %.

One advantage of expressing fractions or ratios as percentages is that their values may then be more readily compared.

For example, if two gun crews have fired practice shoots and the result of the first crew's shooting is 360 points out of a possible 450, while the second crew obtain 615 points out of a possible 750, it is not easy, at a glance, to decide which is the better crew.

When both these results are expressed as percentages, we obtain:

$$\text{First crew's percentage} = \frac{\overset{4}{\cancel{360}}}{\underset{9}{\cancel{450}}} \times \overset{20}{\cancel{100}} = 80\ \%,$$

$$\text{Second crew's percentage} = \frac{\overset{41}{\cancel{615}}}{\underset{15}{\cancel{750}}} \times \overset{2}{\cancel{100}} = 82\ \%,$$

and the comparison is at once evident.

Any percentage may be expressed as a fraction by dividing it by 100, e.g. $37\ \% = \frac{37}{100}$.

When expressing profit from the sale of an article, as a percentage, it must *always* be calculated from the cost, or manufacturing, price and not the selling price.

Example. A firm makes $4\frac{1}{2}d.$ profit by selling an article at 1*s.* 6*d.* What is the percentage profit?

If the article is sold for 1*s.* 6*d.*, and the profit is $4\frac{1}{2}d.$, then the cost price is $1s.\ 6d. - 4\frac{1}{2}d. = 1s.\ 1\frac{1}{2}d.$

So that the percentage profit is

$$\frac{4\frac{1}{2}d.}{1s.\ 1\frac{1}{2}d.} \times 100\ \% = \frac{\cancel{9}}{\cancel{2}} \times \frac{\cancel{2}}{\underset{3}{\cancel{27}}} \times 100\ \% = 33\tfrac{1}{3}\ \%.$$

Exercise XLII.

1. Express these vulgar fractions as percentages:

(*a*) $\frac{1}{2}$. (*b*) $\frac{1}{4}$. (*c*) $\frac{3}{4}$. (*d*) $\frac{1}{8}$. (*e*) $\frac{7}{8}$.

2. Convert these decimal fractions into percentages:

(*a*) 0·18. (*b*) 2·1. (*c*) 0·34. (*d*) 0·7. (*e*) 0·065. (*f*) 2·07.

3. Express the following as percentages to one decimal place:

(*a*) $\frac{1}{3}$. (*b*) $\frac{5}{6}$. (*c*) $\frac{2}{9}$. (*d*) $\frac{3}{13}$. (*e*) $\frac{4}{51}$. (*f*) $\frac{31}{47}$.

4. Express these percentages as vulgar fractions in their lowest terms:

(*a*) 15 %. (*b*) 72 %. (*c*) $6\frac{1}{4}$ %.
(*d*) $102\frac{1}{2}$ %. (*e*) $3\frac{3}{4}$ %. (*f*) 100 %.

5. Convert these percentages into decimal fractions:

(*a*) 16 %. (*b*) $2\frac{1}{2}$ %. (*c*) 125 %.
(*d*) $3\frac{3}{4}$ %. (*e*) 14·2 %. (*f*) 1·05 %.

6. If 1 gallon of Red Sea water weighs 10·40 lb. and 1 gallon of fresh water weighs 9·98 lb., find the percentage of soluble matter (salt, etc.) in the Red Sea.

7. If 1 cubic foot of fresh water weighs 62·36 lb. and increases its volume by 9 % when converted into ice by freezing, find the weight of 1 cubic foot of ice.

8. A floating iceberg has 87 % of its full height submerged. If the height of the iceberg showing above water is 260 ft., at what depth, in fathoms, is the base of the berg below the surface?

9. If the speed of sound in air is 1100 ft. per sec. at a temperature of 5° C., and is 1110 ft. per sec. at a temperature of 10° C., what is the average percentage increase in the speed of sound in air per degree C. between 5° and 10°? (Express the answer as a decimal to two places.)

10. The following relationship holds good, approximately, for any vessel:

$$\frac{\text{Draught of ship in sea water}}{\text{Draught of ship in fresh water}} = \frac{35}{36}.$$

Find (*a*) the percentage increase in draught in changing from sea water to fresh water, (*b*) the percentage decrease in draught in changing from fresh water to sea water. (Answer to nearest second decimal place.)

11. An Admiralty gun-metal contains copper, zinc and tin. If $\frac{1}{2}$ ton of this gun-metal consists of $8\frac{1}{2}$ cwt. of copper, $1\frac{1}{4}$ cwt. of zinc and 28 lb. of tin, what is the percentage composition of this alloy?

12. If the percentage composition of a certain Admiralty gun-metal is 75 % by weight of copper, 15 % by weight of zinc and the remainder tin, find the weight of each metal present in 28 cwt. of this gun-metal.

Propeller Slip.

If a ship's propeller were a perfect form of propulsion, it would mean that one revolution of the propeller, or of the ship's engines that drive it, would move the ship forward a distance equal to the pitch of the propeller blades.

This is never so in actual practice.

The difference between the ship's *theoretical* speed (calculated from propeller or engine revolutions, and the propeller pitch) and the *actual* speed of the ship is known as "propeller slip", and is usually expressed as a percentage of the ship's theoretical speed.

Example. The pitch of a ship's propeller is 12 ft. and the vessel is capable of an actual speed of 21 knots with her engines working at 200 revolutions per min. What is the percentage propeller slip? (Assume 1 N.M. = 2000 yd.)

The theoretical speed of the ship is

$$\frac{12 \times 2\not{0}\not{0} \times \not{6}\not{0}}{\not{6}\not{0}\not{0}\not{0}} \text{ knots} = 24 \text{ knots}.$$

The actual "slip" of the propeller is, therefore, $24 - 21 = 3$ knots.

The percentage propeller slip $= \frac{3}{24} \times 100\ \% = 12\frac{1}{2}\ \%$.

Friction of Pulleys (see p. 63).

The usual allowance for friction in pulley systems is to add, to the weight to be lifted, a percentage of this weight for each moving sheave in the blocks.

This percentage varies according to conditions and to the weight to be lifted, but rarely exceeds $12\frac{1}{2}$ % for each sheave.

EXAMPLE. A man using double Spanish Burton has to lift a load of 5 cwt. What effort, in lb. weight, must he expend, assuming 6 % friction per sheave. (Answer to nearest 10 lb.)

The weight to be lifted is 5 cwt. = 560 lb.

The number of moving sheaves in a double Spanish Burton is 4.

Therefore the allowance for friction, at 6 % per sheave, is

$$24\ \%\ \text{of}\ 560\ \text{lb.} = \tfrac{24}{100} \times 560\ \text{lb.} = 134{\cdot}4\ \text{lb.}$$

Thus the equivalent weight to be lifted is

$$560 + 134{\cdot}4\ \text{lb.} = 694{\cdot}4\ \text{lb.}$$

The effort-load ratio of double Spanish Burton is 1:5.

Therefore the effort required is $\dfrac{694{\cdot}4}{5}$ lb. = 138·9 lb.

The required effort, to the nearest 10 lb., is therefore 140 lb.

EXERCISE XLIII.

Assume, where necessary, 1 N.M. = 2000 yd.

1. The pitch of a ship's propeller is 14 ft. and the vessel is capable of a speed of 18 knots when her engines are developing 150 revolutions per min. What is the percentage propeller slip? (Answer to nearest first decimal place.)

2. The theoretical speed of a ship is 24·2 knots. Her maximum speed is 22 knots. What is her percentage propeller slip? (Answer to nearest first decimal place.)

3. A vessel's maximum speed is 27 knots and her propeller slip is 10 %. What is her theoretical speed?

4. A vessel's propeller slip is $13\frac{1}{3}$ % and while revolving at 150 revolutions per min. would give the ship an actual speed of 13 knots. What is the pitch of the propeller?

5. A man using double luff tackle has to lift a load of 350 lb. If the allowance for friction is 5 % per sheave, what effort must he use?

6. A man using single Spanish Burton with an allowance of 10 % per sheave for friction exerts a pull of 120 lb. What is the maximum load he can lift?

7. The safe working load of a rope is 3 cwt. A man uses this rope with ordinary luff tackle and attempts to lift a load of 6 cwt. If the allowance for friction is 10 % per sheave, will the rope stand the strain, and if so by how many lb. wt.? (Answer to nearest lb.)

8. A rope whose normal breaking strain is 1 ton and which has been spliced in one place is used with threefold purchase to lift a load of 5 cwt. The allowance for friction is 6 % per sheave. Is the safe working load of the rope exceeded, and if so by how many lb. wt.? (Answer to nearest lb.)

ACCOUNTS

The keeping of simple accounts should be within the province of every rating in a ship.

First, for his own personal benefit, he should be able to calculate what pay is due to him. This is his own PRIVATE PAY ACCOUNT.

Secondly, a rating should be able and willing to assist the smooth running of his mess by acting as mess caterer, or messman, and should be able to issue for the benefit of his messmates a written balanced statement of monies received and monies expended. This is called a MONTHLY MESS ACCOUNT.

Thirdly, when acting as messman, a rating should be able to check all purchases made by keeping a MESSMAN'S DETAILED STATEMENT.

Pay Accounts.

Boys and men in the Royal Navy are paid their "wages" at a daily rate, which is known as the basic daily rate and applies to every day in the year.

In addition to his basic daily rate of pay a rating may receive extra payment either as a daily allowance, a period allowance, or a lump sum payment.

These extra payments may be any or all of the following:

(*a*) Kit upkeep allowance for renewal of uniform.

(*b*) Good conduct pay, granted upon issue of good conduct (G/C) badges.

(*c*) Grog money, in lieu of grog (if over 20 years of age).

(*d*) Extra pay as specialists in gunnery, torpedo work, wireless telegraphy, etc.

(*e*) Leave allowance for victuals if on long leave.

(*f*) Marriage allowance if married and over 20 years of age.

(*g*) Special service allowance, while serving in small craft, submarines, etc.

A rating may, and usually does, allot a weekly sum of money to his parents or his wife and this is stopped out of his pay and paid directly to them.

Deductions are also made for clothing items ("slops") drawn from stores, tobacco, etc., and also for insurance contributions.

Men in the Navy are paid at intervals, usually fortnightly when at home, or monthly when abroad. In either case a *percentage* only of the amount due is paid to a man, so that he is left with a credit in the ship's ledgers.

At the end of each quarter of the year, i.e. on 31 March, 30 June, 30 September and 31 December, a man draws the total of pay and allowances due to him, after deduction of all payments made on his account, at what is called the "quarterly pay down".

The number and variety of the allowances to which a man may be entitled make the calculation of his quarterly pay down his own peculiar problem.

EXAMPLE. A Petty Officer receives 6*s*. 5*d*. per day as pay and a war bonus of 1*s*. 0*d*. per day. As gunner's mate he is entitled to an extra 2*s*. 0*d*. per day. He has two G/C badges, for each of which he is granted 3*d*. per day. He has been credited with 12 days' leave allowance at the rate of 2*s*. 2*d*. per day, 13 weeks' grog allowance, in lieu of grog, at 1*s*. 7*d*. per week and £1. 15*s*. 0*d*. kit upkeep allowance.

During the quarter he has drawn pay amounting to £16 and his quarterly allotment to his wife is £22. His "stoppages" during the quarter also include 4*s*. 3*d*. for insurance and 17*s*. 3*d*. for "slops" drawn from the Paymaster's stores.

In a quarter consisting of 92 days, find what payment is due to him at the quarterly pay down.

His total daily rate of pay is

6*s*. 5*d*. + 1*s*. 0*d*. + 2*s*. 0*d*. + 6*d*. = 9*s*. 11*d*. per day.

Therefore the total pay due to him for a quarter of 92 days is

92 × 9*s*. 11*d*. = £45. 12*s*. 4*d*.

His remaining allowances are

12 × 2*s*. 2*d*. + 13 × 1*s*. 7*d*. + £1. 15*s*. 0*d*.
= £1. 6*s*. 0*d*. + £1. 0*s*. 7*d*. + £1. 15*s*. 0*d*.

His pay account, therefore, should be drawn up in this way.

Credit	£	*s*.	*d*.	Debit	£	*s*.	d.
Pay at 9*s*. 11*d*. per day	45	12	4	Allotment to wife	22	0	0
				Pay received	16	0	0
Leave allowance at 2*s*. 2*d*. per day	1	6	0	Slops		17	3
				Insurance		4	3
Grog money	1	0	7				
Kit upkeep	1	15	0				
Credit total	49	13	11	Debit total	39	1	6

Pay down = £10. 12*s*. 5*d*.

EXERCISE XLIV.

1. A shipwright (5th class) received 5*s.* 7*d.* per day and 6*d.* per day war bonus between 1 January 1940 and 31 March 1942. What pay did he receive (*a*) for the year 1940, (*b*) for the period 1 January 1942 to 31 March 1942?

2. If the pay of a blacksmith (4th class) including war bonus during 1941 was £123. 3*s.* 9*d.* for the whole year, find (*a*) his daily rate of pay, including war bonus, (*b*) the payment he received for 142 days' service.

3. The pay of a certain rating is 4*s.* 9*d.* per day with an additional 1*s.* 0*d.* per day war bonus. He receives 3*d.* per day for each of his two G/C badges and 1*s.* 7*d.* per week grog allowance. What should he receive for a quarter consisting of 91 days?

4. The pay of a Petty Officer was 7*s.* 6*d.* per day from 1 January 1940 to 10 February 1940 inclusive. His pay was then increased to 7*s.* 9*d.* per day. He was also entitled to 6*d.* per day war bonus, 1*s.* 7*d.* per week grog allowance, and £1. 12*s.* 6*d.* kit upkeep allowance. What was his total credit for the quarter ending 31 March 1940?

5. The rate of pay of a Petty Officer, including war bonus, was 9*s.* 0*d.* per day from 1 October to 21 November inclusive. On 22 November he was granted a G/C badge. His kit upkeep allowance was £1. 17*s.* 0*d.* and grog allowance £1. 0*s.* 7*d.* His stoppages were as follows: (*a*) pay advanced, £14; (*b*) allotment to mother, £16; (*c*) slops drawn, 10*s.* 7*d.*; (*d*) insurance contribution, 5*s.* 10*d.* Draw up his pay account for the quarter ending 31 December and show what his pay down should be in dollars if he is paid in New York with the American dollar equivalent to 4*s.* 2*d.* in English money.

6. Draw up the pay account of a Petty Officer from the following details and show his quarterly pay down: rate of pay, 7*s.* 3*d.* per day; war bonus, 6*d.* per day; 3 G/C badges at 3*d.* per day each; 2*s.* 0*d.* per day as torpedo gunner's mate; kit upkeep allowance, £1. 16*s.* 0*d.*; 14 days' leave allowance at 2*s.* 2*d.* per day; grog allowance, 1*s.* 7*d.* per week. Pay drawn, £21, allotment to wife, £20, slops from steward to value of £1. 3*s.* 6*d.* Date of quarterly pay down 31 March 1940.

7. A Petty Officer with three G/C badges has a basic rate of pay of 8*s.* 0*d.* per day on 1 July 1941, exclusive of his 6*d.* per day war bonus. On 3 August he receives the rating of gunner's mate with an increase of pay of 2*s.* 0*d.* per day. On and from 3 August he increases the allotment to his wife from 4*s.* 0*d.* per day to 5*s.* 0*d.* per day. His total quarterly allowances come to £2. 15*s.* 6*d.* and he has drawn slops to the value of £1. 4*s.* 8*d.* If his pay already received amounts to £19. 0*s.* 0*d.*, draw up his pay account for 30 September 1941 and show what pay down he should receive.

8. On 1 April a Petty Officer is receiving 7*s.* 3*d.* per day basic pay and has 2 G/C badges. His war bonus is 6*d.* per day and his extra pay as a leading torpedoman is 1*s.* 6*d.* per day. On 20 May he goes on leave for

14 days, during which he is credited with 2*s*. 2*d*. per day leave allowance. He also draws 1*s*. 7*d*. per week grog allowance. His allotment to his wife is £2. 0*s*. 0*d*. per week and his total of pay drawn since 31 March is £7. 10*s*. 0*d*. On and from 5 June he is rated torpedo gunner's mate with an increase of pay of 6*d*. per day, and has orders to join a new ship by 6 a.m. on 10 June. What credit balance should be transferred in his name to the new ship?

Mess Accounts.

Every man in the Royal Navy is supplied with all his meals. For this purpose men are grouped in convenient numbers, and with due regard to their work, into what are called "messes".

Every man in the Service is entitled to a daily "victualling allowance" credited to him in money, and some particular member of the mess is elected mess caterer, or messman, and uses the monthly total of these monies to cater for the food of his mess.

A settlement of mess accounts is made, usually at the end of every calendar month, when the messman draws up his statement of accounts.

This is not quite such a straightforward proceeding as might be expected at first sight, because the numbers victualled in a mess are frequently changing. Men may be transferred; new men may join; some may go on leave, and so on. The method adopted for the administration of victualling funds is the same for all messes and the procedure is as follows.

The actual money due to the mess from victualling funds is held in credit by the ship's Paymaster. For example, a mess of 20 men, with a victualling allowance of 1*s*. 1*d*. per head per day, for a month of 30 days, would be credited with a messing allowance, for that month, of

$$20 \times 30 \times 1s.\ 1d. = £32.\ 10s.\ 0d.$$

The mess is often entitled to further allowances, such as special dinner vouchers, etc., and the total sum to which the mess is entitled is credited to them by the Paymaster.

The messman orders certain standard items of provisions, such as bread, meat, etc., from the victualling department, which is controlled by the Paymaster. These items are signed for by the messman on behalf of the mess, and the Paymaster pays for them out of the mess funds. These items are known as "provisions issued on repayment".

Any extra supplies, other than the standard items, are ordered by the messman from the ship's canteen. These, also, are paid for by the Paymaster from the mess funds and come under the heading of "canteen bill"

Any replacements, due to loss or breakage, of articles of cutlery, crockery, etc. (called "mess traps") are also paid for by the Paymaster out of the mess funds.

So that at the end of the month the Paymaster deducts from the messing allowance due to the mess the following three items:

(i) provisions issued on repayment,
(ii) canteen bill,
(iii) mess traps replaced.

Any balance left over is then paid to the mess by the Paymaster, or any deficit must be paid for by the members of the mess to get it out of debt.

The messman, therefore, has to make out two separate accounts,

(i) the account of mess with Paymaster,

(ii) mess statement.

The latter account must include any additional expenditure or purchases by the mess which are not administered by the Paymaster, and may include monies due from individual members to the messman for the supply of items which have been purchased privately.

Both statements are drawn up in two parallel columns, one for "income" and the other for "expenditure". These columns must be made to "balance", either by showing the mess to have money in hand, or by contributions from each member of the mess to make up the deficiency, if the mess is in debt.

EXAMPLE. The following is a typical mess account.

Mess "A" for March 1942. Number of men in mess: 21.

	£	s.	d.
Messing allowance	23	14	0
Special dinner vouchers		10	8
Provisions issued on repayment	12	13	5
Mess traps lost		7	10
Canteen bill	10	12	4
Extra expenses during month	1	18	7
Due from messmates for private supplies		12	2

Find (*a*) the amount due to the mess from the Paymaster, (*b*) the amount each member of the mess must pay to keep the mess out of debt.

MESS ACCOUNT. March 1942

Account of mess with Paymaster. No. in mess: 21

Income	£	s.	d.	Expenditure	£	s.	d.
Messing allowance	23	14	0	Provisions issued on repayment	12	13	5
Special dinner vouchers		10	8	Lost mess traps		7	10
				Canteen bill	10	12	4
				Due to mess from Paymaster		11	1
Total	24	4	8	Total	24	4	8

Mess statement

Income				Expenditure			
	£	s.	d.		£	s.	d.
Due to mess from Paymaster		11	1	Extra expenses	1	18	7
Private supplies to messmates		12	2				
Contributions (21 at 9*d.*)		15	9				
				Balance in hand			5
Total	1	19	0	Total	1	19	0

The last entry in the "income" column of the mess statement is obtained by finding the total of the receipts, i.e. 11*s.* 1*d.* + 12*s.* 2*d.* = £1. 3*s.* 3*d.* This is subtracted from the expenditure, i.e. £1. 18*s.* 7*d.* leaving a deficiency of 15*s.* 4*d.* This is divided by 21 to find the amount each member must contribute to pay off the debt, i.e.

$$\frac{15s.\ 4d.}{21}=\frac{184d.}{21}=8\tfrac{16}{21}d.$$

Therefore the smallest payment that will clear the debt is 9*d.* per head = 21 × 9*d.* = 15*s.* 9*d.*, leaving a balance in hand in the mess funds of 5*d.*

The answers therefore are (*a*) 11*s.* 1*d.*, (*b*) 9*d.* per head, leaving a balance of 5*d.* in the mess funds.

Messman's Detailed Statements.

The extra expenses incurred by the messman on behalf of the mess must, for the benefit of the rest of the mess, be shown as a detailed statement.

EXAMPLE. During one month a messman made the following purchases ashore:

1 cwt. new potatoes at $1\frac{1}{2}d.$ per lb.; 3 dozen cabbages at $2\frac{1}{2}d.$ each.; 12 tins of salmon at 1*s.* 6*d.* per tin; 12 lb. of oranges at $7\frac{1}{2}d.$ per lb.; 20 lb. of tomatoes at 1*s.* $2\frac{1}{2}d.$ per lb. Railway and bus fares ashore cost him 6*s.* 10*d.*

Then (*a*) Find the total cost of the extra purchases. (*b*) If there were 30 men in the mess and the messman had already taken up stores in excess of his mess allowance due from the Paymaster by £2. 7*s.* 8*d.*, find to the nearest penny the contribution from each man in the mess to cover the total deficit.

Messman's statement for month. Mess "B"

	£	s.	d.
1 cwt. new potatoes at $1\frac{1}{2}d.$ per lb.		14	0
3 dozen cabbages at $2\frac{1}{2}d.$ each		7	6
12 tins of salmon at 1*s.* 6*d.* per tin		18	0
12 lb. of oranges at $7\frac{1}{2}d.$ per lb.		7	6
20 lb. of tomatoes at 1*s.* $2\frac{1}{2}d.$ per lb.	1	4	2
Messman's fares and expenses		6	10
Total of extra purchases	3	18	0
Debt to Paymaster	2	7	8
Total	£6	5	8

Contribution from each member

$$=\frac{£6.\ 5s.\ 8d.}{30}=4s.\ 2\tfrac{8}{30}d.=4s.\ 3d. \text{ to nearest penny}$$

necessary to clear the debt.

The answers are, therefore, (*a*) £3. 18*s.* 0*d.*, (*b*) 4*s.* 3*d.*

EXERCISE XLV.

1. Mess account for July 1942. 21 men in mess.

	£	s.	d.
Messing allowance	24	15	0
Special dinner vouchers		12	6
Provisions issued on repayment	13	13	6
Mess traps lost		5	9
Canteen bill	11	0	3
Extra expenses during month	1	15	4
Due from messmates for private supplies		17	10

Find (*a*) the amount due to the mess from the Paymaster, (*b*) the amount each man must pay, to the nearest penny, to keep the mess out of debt.

2. Mess account for April 1942. 18 men in mess.

	£	s.	d.
Messing allowance	20	10	6
Special dinner vouchers		9	0
Provisions issued on repayment	9	1	2
Mess traps replaced		5	8
Canteen bill	13	8	2
Extra expenses during month	2	11	0
Due from messmates for supplies		5	6

Find (*a*) the amount due from the mess to the Paymaster, (*b*) the amount each man must pay to clear the debt.

3. Mess account for September 1942. 25 men in mess.

	£	s.	d.
Messing allowance	25	12	6
Special dinner vouchers		11	2
Provisions issued on repayment	14	9	3
Mess traps lost		10	8
Canteen bill	11	7	8
Extra expenses during month	2	5	6
Due from messmates for supplies		12	6
Balance in mess funds from August	2	12	0

Find (*a*) the amount due from the mess to the Paymaster, (*b*) balance carried forward to October.

4. Mess account for September 1942. 22 men in mess.

	£	s.	d.
Messing allowance	26	2	0
Special dinner vouchers		13	6
Provisions issued on repayment	14	4	6
Mess traps replaced		7	3
Canteen bill	11	11	3
Extra expenses during month	2	10	8
Private supplies to messmates		17	10
Balance in hand from August		5	8

Find (*a*) the amount due to the mess from the Paymaster, (*b*) the amount each man must pay to free the mess from debt.

5. Mess account for July 1942. 24 men in mess.

	£	s.	d.
Messing allowance	24	7	4
Special dinner vouchers		11	8
Provisions issued on repayment	14	2	3
Mess traps lost		9	3
Canteen bill	10	14	6
Extra expenses during month	2	1	9
Private supplies to messmates		15	7

Find (*a*) the amount due from the mess to the Paymaster, (*b*) the amount each man must pay to clear all debt and leave a balance in hand of at least ten shillings.

6. During a month a messman exceeded his allowance from the Paymaster by £1. 15*s.* 10*d.* and in addition made the following outside purchases: fresh fruit, 17*s.* 10*d.*; 65 lb. of fish at 8½*d.* per lb.; 34 lb. of onions at 2½*d.* per lb.; 3 legs of lamb each weighing 6 lb. at 1*s.* 4*d.* per lb. His personal fares and expenses ashore were 6*s.* 4*d.* Find the total debt of the mess, allowing a discount on outside purchases of 6*d.* on every complete £1.

7. A mess of 45 Petty Officers gave a children's party and £25 was voted from the mess funds for this purpose. Any cost over and above £25 was to be met by a levy on all members of the mess to the nearest shilling to cover the deficiency. The expenses incurred were as follows: 5 lb. of tea at 1*s*. 10½*d*. per lb.; 9 lb. of sugar at 3½*d*. per lb.; 12 lb. of margarine at 10½*d*. per lb.; 14 lb. of cake at 9½*d*. per lb. Bread and sundries cost £2. 10*s*. 6*d*. Hire of band and hall £17. 10*s*. 0*d*. Presents for the children £15. 0*s*. 0*d*. What credit was returned to the mess funds?

8. During September a messman took up stores, in excess of his allowance due from the Paymaster, by £3. 13*s*. 4*d*. In addition his outside purchases were: 1½ cwt. of potatoes at 2¼*d*. per lb.; 45 lb. of fresh fish at 10½*d*. per lb.; 6 dozen oranges at 2 for 3½*d*.; 2½ dozen cabbages at 2½*d*. each. His fares and expenses ashore were 7*s*. 10*d*. and he obtained a discount of 1*s*. 0*d*. in the £1 on every complete £ of outside purchases. If there were 40 men in the mess, find, to the nearest penny, each member's contribution to cover the total debt.

9. A mess, having a monthly account with a stores ashore, has been supplied during the month with the following goods: 48 lb. of fresh haddock at 8½*d*. per lb.; 30 lb. of plaice at 1*s*. 3*d*. per lb.; 36 lb. of Torbay sole at 1*s*. 2½*d*. per lb.; 24 lb. of hake at 7½*d*. per lb.; 72 herrings at 3 for 4*d*. If the mess is allowed discount at the rate of 7½*d*. in the £1 for every complete £, what is the cost debited to the mess?

10. A mess of 50 Chief Petty Officers gave a children's party and £30 was voted from the mess funds towards the cost. The remainder was to be met by a levy on all members to the nearest 6*d*. The following expenses were incurred: hall and band, £21. 10*s*. 0*d*.; children's presents, £16. 5*s*. 0*d*.; 6½ lb. of tea at 1*s*. 9½*d*. per lb.; 10 lb. of sugar at 3½*d*. per lb.; 16 lb. of cake at 11½*d*. per lb.; 16 lb. of margarine at 1*s*. 1*d*. per lb.; bread and sundries £2. 8*s*. 6*d*. What credit was returned to mess funds and how much did each member contribute?

SQUARE ROOT

Many interesting and important calculations involve the use of square roots. Visibility distance at sea, distances from which the lights of lightships and lighthouses can be seen, range finding, calculation of sail areas, etc. are a few of such problems.

It is necessary therefore to know what is meant by the square root of a number and how it is obtained.

Square Root.

When a number is multiplied by itself it is said to be "squared": e.g. 8 squared is $8 \times 8 = 64$, and 64 is the square of 8.

This is written thus: $8^2 = 64$.

The reverse of squaring is to find the "square root": e.g. the square root of 64 is 8.

This is written thus: $\sqrt{64}=8$.

The square of the *product* of two numbers is the *product* of their squares: e.g.

$$(5\times 3)^2=5^2\times 3^2=25\times 9=225.$$

Similarly, the square root of the *product* of two numbers is the *product* of their square roots: e.g.

$$\sqrt{121\times 49}=\sqrt{11\times 11\times 7\times 7}=11\times 7=77.$$

Similarly with fractions, thus

$$\sqrt{\frac{25}{64}}=\sqrt{\frac{5\times 5}{8\times 8}}=\frac{5}{8}.$$

By ordinary multiplication we can draw up a table of the squares of the numbers from 1 to 20:

Number	Square	Number	Square
1	1	11	121
2	4	12	144
3	9	13	169
4	16	14	196
5	25	15	225
6	36	16	256
7	49	17	289
8	64	18	324
9	81	19	361
10	100	20	400

It is at once clear from this table that although every number must have its corresponding square, it by no means follows that every number must have an *exact* square root.

For example, from the table above, the square root of 9 is 3 and the square root of 16 is 4, i.e. $\sqrt{9}=3$, and $\sqrt{16}=4$.

Therefore the square root of any number between 9 and 16 must be greater than 3 and less than 4.

Example. What is the square root of 13?

Let us write 13 as an improper fraction $=\frac{325}{25}$. And from the table above we see that $\sqrt{324}=18$, so that $\sqrt{325}$ is just a very little greater than 18. Thus

$$\sqrt{13}=\sqrt{\frac{325}{25}}=\frac{18}{5}\text{ nearly}=3\cdot 6\text{ nearly}.$$

A still closer approximation is 3·606.

The arithmetical method used to determine the square root of a number may, if required, be obtained from any arithmetical text-book, but for purposes of speed and simplicity it is more convenient for us to use a table of square roots, as provided at the back of the book.

Reading of Square Root Tables.

The method of "reading" a table of square roots may be explained in this way.

First of all, write down, in table form, the squares of the following numbers: 0·06, 0·6, 6, 60, i.e.

$$0{\cdot}06^2 = 0{\cdot}0036,$$
$$0{\cdot}6^2 = 0{\cdot}36,$$
$$6^2 = 36,$$
$$60^2 = 3600.$$

Thus

$$\sqrt{0{\cdot}0036} = 0{\cdot}06,$$
$$\sqrt{0{\cdot}36} = 0{\cdot}6,$$
$$\sqrt{36} = 6,$$
$$\sqrt{3600} = 60.$$

What is the square root of 0·036, 3·6 and 360?

From the above table of squares of numbers, we see that $19^2 = 361$, i.e. $\sqrt{360} = 19$ nearly.

Also we know $1{\cdot}9^2 = 3{\cdot}61$, i.e. $\sqrt{3{\cdot}6} = 1{\cdot}9$ nearly, and $0{\cdot}19^2 = 0{\cdot}0361$, i.e. $\sqrt{0{\cdot}036} = 0{\cdot}19$ nearly.

So that, if we now write these square roots in order of magnitude, we have

$$\sqrt{0{\cdot}0036} = 0{\cdot}06,$$
$$\sqrt{0{\cdot}036} = 0{\cdot}19 \text{ nearly},$$
$$\sqrt{0{\cdot}36} = 0{\cdot}6,$$
$$\sqrt{3{\cdot}6} = 1{\cdot}9 \text{ nearly},$$
$$\sqrt{36} = 6,$$
$$\sqrt{360} = 19 \text{ nearly},$$
$$\sqrt{3600} = 60.$$

Consequently we must allow for this peculiarity by which the same given *figures* of a number may have two quite different sets of figures for the square root.

If we examine the table of square roots at the end of the book, we see that the number 36 in the left-hand column has two groups of four-figure numbers adjacent to it, viz. 1897 and 6000. To determine which group of figures to use for the correct square root we proceed in this way.

Suppose that we wish to find the square root of 366·2.

By reference to our column of square roots above we see that when the number under the square root sign has

1 or **2** figures in the whole number part, the square root has **1** figure in the whole number part, and

3 or **4** figures in the whole number part, the square root has **2** figures in the whole number part, and so on.

Thus the square root of 366·2 has 2 figures in the whole number part.

Now mark off the number 366·2 into pairs of figures to the left of the decimal point, thus 3|66|·2.

The figure in the last group, to the left, is 3, and the square root of 3 is greater than 1 but less than 2. So that the square root of 366·2 must commence with 1. . . . .

Referring now to the table of square roots, we first find the number 36 in the left-hand column and select the group of figures adjacent which commence with 1. . . . . This is 1897. Following along this line we come to the number 1913 beneath the figure 6 in the top line. In the three smaller columns to the right (called the "difference columns") we find, under the figure 2 in the top line and in the same line as the number 1913, the figure 1.

This must be added to 1913 to obtain the figures in the answer, i.e. 1914.

By placing the decimal point so that there are 2 figures in the whole number portion, the square root of 366·2 is 19·14.

Suppose now we wish to find $\sqrt{0{\cdot}003662}$.

This time, since there are no whole numbers, we mark off pairs of figures to the right of the decimal point: 0·|00|36|62|.

For every *pair* of noughts after the decimal point in the original number there is *one* nought after the decimal point in the square root.

The square root of the first pair of figures containing a significant figure, i.e. $\sqrt{36}$, is 6. So that the required square root must commence with the figures 0·06. . . .

By reference to the tables of square roots as before, we find the square root of 0·003662 is 0·06052. As will have been noticed, the tables of square roots cannot be used for numbers containing more than four significant figures, so we have to approximate for numbers containing more than four figures.

Thus $\sqrt{321{\cdot}846}$ is approximately $\sqrt{321{\cdot}8}$ and $\sqrt{0{\cdot}200392}$ is approximately $\sqrt{0{\cdot}2004}$.

EXERCISE XLVI.

1. Write down the whole number portion of these square roots by inspection:

(*a*) $\sqrt{5}$. (*b*) $\sqrt{18}$. (*c*) $\sqrt{37}$. (*d*) $\sqrt{68}$. (*e*) $\sqrt{99}$.

2. From the tables find the square root of:

(*a*) 841. (*b*) 1521. (*c*) 9409. (*d*) 1444. (*e*) 8649.

3. From the tables find to four significant figures the square root of:

(*a*) 32·49. (*b*) 3·249. (*c*) 176·4. (*d*) 7·335.
(*e*) 0·01621. (*f*) 1·025. (*g*) 0·0003162. (*h*) 9·006.

4. Find, from the tables, the approximate square root of:

(*a*) 174·388. (*b*) 2·9117. (*c*) 0·039491. (*d*) 3·00686.
(*e*) 11·0622. (*f*) 0·030303. (*g*) 2973·69. (*h*) 11,033.

Visibility Distance at Sea.

Because of the curvature of the earth's surface, the distance of visibility, on a clear day, depends upon the height of the observer's eye above sea level. In other words, the distance of the horizon varies according to the height of the observer.

In practice, this distance is estimated by using a simple formula:

$$D=1{\cdot}15\sqrt{h},$$

where D is the horizon distance in nautical miles and h is the height in feet of the observer's eye above sea level.

Thus, at a height of 36 ft., the distance of visibility to the horizon is

$$D=1{\cdot}15\sqrt{36}=1{\cdot}15\times 6=6{\cdot}9 \text{ nautical miles.}$$

EXAMPLE 1. At what height must a light be placed so as to be visible from a distance of 8 N.M. to an eye at sea level?

Using the above formula, we have

$$D=1{\cdot}15\sqrt{h}, \quad \text{i.e. } 8=1{\cdot}15\sqrt{h}.$$

So that $$\sqrt{h}=\frac{8}{1{\cdot}15}=6{\cdot}95 \text{ ft.}$$

Therefore $$h=(6{\cdot}95)^2=48{\cdot}3 \text{ ft.}$$

EXAMPLE 2. If an observer is 40 ft. above sea level and a light is 50 ft. above sea level, what is the greatest distance of visibility of the light?

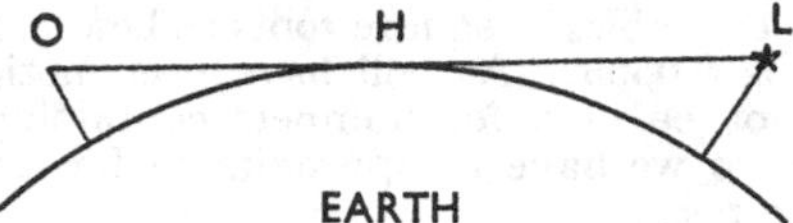

In the accompanying figure, let L be the light and O the observer. Then the distance of visibility from O to the horizon H is the line OH. And the distance of visibility from L to the horizon H is the line LH.

Therefore the greatest distance of visibility of the light L from O is $OH+HL$.

So that greatest distance

$$=1{\cdot}15\sqrt{40}+1{\cdot}15\sqrt{50}=1{\cdot}15\times 6{\cdot}325+1{\cdot}15\times 7{\cdot}07\dot{1}$$

$$=7{\cdot}27375+8{\cdot}13165=15{\cdot}4 \text{ N.M. (approximately).}$$

The height of any object, when shown on a chart, is given always as its height above water at high-water spring tide. That is to say, the height above the sea of any charted object can never be *less* than that shown on the chart.

Therefore it follows that the range of visibility of a charted light is never less than what it is for its charted height.

When the distance of visibility of a light is charted, it is given as that at H.W.S.T. for an observer whose height of eye is 15 ft. above sea level.

EXERCISE XLVII.

1. What is the distance to the horizon from an observer whose eye level above the sea is:

(*a*) 10 ft. (*b*) 45 ft. (*c*) 90 ft. (*d*) 110 ft.?

2. Find, to the nearest foot, the height above sea level of lights which are visible to an observer at sea level at distances of:

(*a*) 15 N.M. (*b*) 21 N.M. (*c*) 30 N.M.

3. A light is charted as 120 ft. What is its distance of visibility to an eye at sea level, (*a*) at H.W.S.T., (*b*) when the tide is 12 ft. below H.W.S.T.? (Answer to nearest first decimal place.)

4. A light charted as 134 ft. is just visible to an observer at sea level at a distance of 13·8 N.M. How many feet is the sea level below H.W.S.T.?

5. St Catherine's Light is charted as 136 ft. What should its visibility distance be charted, to the nearest nautical mile?

6. A look out in a crow's-nest, 72 ft. above sea level, can just see a light from a lighthouse charted as 174 ft. The tide is known to be 8 ft. below H.W.S.T. How far, to the nearest nautical mile, is the ship from the light?

7. A light is charted as having a range of visibility of 19·4 N.M. What should be its charted height to the nearest foot?

8. A light which is charted as having a range of visibility of 14·8 N.M. is just visible from sea level at a distance of 11·09 N.M. Find, to the nearest foot, how many feet the sea level is below H.W.S.T.

9. A light is charted as having a range of visibility of 16 N.M. If the tide is 15 ft. below H.W.S.T., what should be the visibility range of the light to an observer at sea level? (Answer in nautical miles to the nearest first decimal place.)

10. A light is charted as having a range of visibility of 21·7 N.M. If the tide is 11 ft. below H.W.S.T., what should be the visibility range of the light to an observer at a height of 25 ft. above sea level? (Answer in nautical miles to the nearest first decimal place.)

MENSURATION

Areas of Plane Surfaces.

Theoretically, a plane (or flat) surface has length and breadth but no thickness, but for purposes of practical measurement all materials dealt with in flat sheets, such as sails, tarpaulins, paper, etc., are considered to be plane surfaces.

The shape of a plane surface is defined by the shape of its edges, e.g. an untorn sheet of paper is usually a rectangle or a square.

Regular Four-sided Figures.

(1) A *Parallelogram* is a plane four-sided figure with its opposite sides equal and parallel (Fig. *a*).

(2) A *Rectangle* is a parallelogram with all its angles right angles (Fig. *b*).

(3) A *Square* is a rectangle with all its sides equal (Fig. *c*).

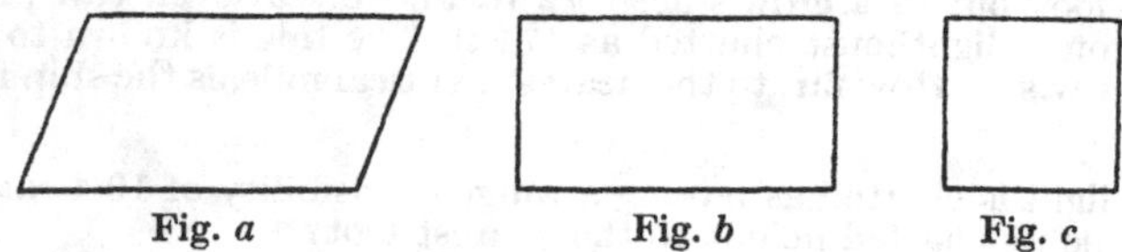

Fig. *a* Fig. *b* Fig. *c*

If on the same base *AB* we draw a rectangle *ABCD* and a parallelogram *ABEF* (Fig. *d*), we can show that the area *ADF* is equal to the area *BCE*. So that the area of the rectangle *ABCD* is equal to the area of the parallelogram *ABEF*.

The area of a plane figure is expressed in the *square* of the units in which its sides are measured.

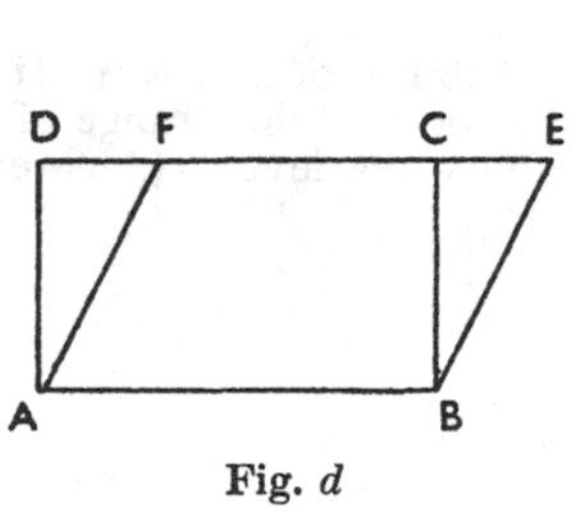

Fig. *d*

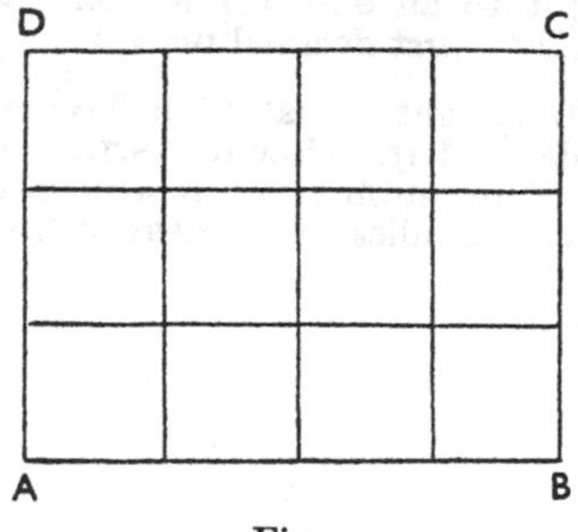

Fig. *e*

Suppose *ABCD* (Fig. *e*) to be a rectangle with sides *AB* and *CD* each = 4 in., and sides *BC* and *DA* each = 3 in.

If we then divide the rectangle into squares each of 1 in. sides, we have 12 of such squares making up the rectangle.

In other words, the area of the rectangle *ABCD*

$$= 12 \text{ inch squares} = 12 \text{ square inches}$$
$$= 4 \times 3 \text{ square inches.}$$

So that the area of a rectangle, in square units, is equal to the product of the lengths of two of its adjacent sides.

A rectangle whose adjacent sides are 3·1 in. and 2·7 in. in length has an area of $3{\cdot}1 \times 2{\cdot}7$ sq. in. $= 8{\cdot}37$ sq. in.

The area of a square is obviously equal to the square of its side, i.e.

Area of a square of side 3·3 cm. $= 3{\cdot}3^2$ sq. cm. $= 10{\cdot}89$ sq. cm.

EXERCISE XLVIII.

1. What is the area of a rectangular cabin floor measuring 10 ft. by 8 ft. 6 in.?

2. What area of carpet, in square yards, would be needed to cover the floor in question 1, leaving a margin 6 in. wide all round?

3. A certain red lead paint will cover at the rate of 450 sq. ft. per gallon for a first coat, and 600 sq. ft. per gallon for each succeeding coat. How many gallons of this paint will be required to paint, with three coats, both sides of a bulkhead measuring 20 yd. by 5 yd.?

4. One pound of stiff white lead paint when thinned will cover $7\frac{1}{2}$ sq. yd. of smooth surface. How many pounds would be needed to paint two coats on the outside of all six smooth sides of a rectangular water tank measuring 13 ft. 6 in. long, by 7 ft. 6 in. wide, by 6 ft. deep? (Answer to nearest lb.)

5. A ship's funnel is 50 ft. high and has a circumference of 62 ft. 6 in. How much black paint, to the nearest lb., would be needed to paint this with two coats if the paint covers 8 sq. yd. per lb.?

6. What is the area in square yards of two white rings painted on the funnel in question 5, both rings being 4 ft. 6 in. wide?

7. A hatch on an upper deck leading to the hold measures 24 ft. by 12 ft. It is closed by planks of wood 12 ft. long by 16 in. wide. How many of these planks are required to close the hatch?

8. To prepare for sea, a hatch measuring 24 ft. by 12 ft. is battened down and covered with a tarpaulin which overhangs the hatch combings by 18 in. all round. What is the area of this tarpaulin in square yards?

9. A sailmaker is ordered to make an upper deck awning to measure 58 ft. 6 in. by 12 yd., and to be edged with rope. The canvas is supplied in "bolts" of 39 yd. long by 2 ft. wide, weighing 35 lb. per bolt. Find (*a*) the area of the awning in square feet, (*b*) the length of rope in fathoms required for edging, (*c*) the weight of canvas needed. (Neglect any allowance for seams.)

10. A tarpaulin measuring 9·5 metres by 6 metres is to be cut to cover a hatch measuring 27 ft. by 15 ft. 6 in. and allow a margin of overhang of 18 in. all round. How many square feet will be left over? (Answer to nearest square foot, assuming 1 metre = 3·28 ft.)

Triangles.

A triangle is a plane figure bounded by three straight lines.

The area of a triangle may be obtained in two ways: (i) geometrical method, (ii) arithmetical method.

Geometrical method. Just as we know that the area of a rectangle or a parallelogram is equal to the product of the base and perpendicular height, so the area of any triangle is *half* the product of the base and perpendicular height, e.g. the area of the triangle ABC (Fig. *f*) is $\frac{1}{2}AB \times CD$.

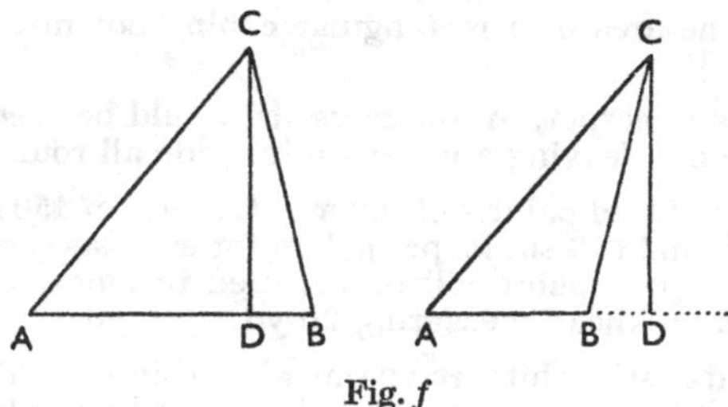

Fig. *f*

Unfortunately, in practice, the perpendicular height of a triangular area is seldom given in its dimensions. Instead, we usually know the length of each of its three sides. Consequently the perpendicular height has to be determined by scale drawing. This depends for its accuracy almost entirely upon careful measurement of straight lines, and for this purpose a *diagonal scale* should be used. This can be purchased or one may be drawn upon good quality card.

A diagonal scale may be constructed as follows:

With a ruler and set square draw a rectangle 4 in. long by 1 in. high, and draw perpendiculars at 1 in. intervals as shown in Fig. *g*.

This will divide the rectangle into four 1 in. squares.

Divide one of these perpendiculars into 10 equal parts (i.e. each division is $\frac{1}{10}$ in.).

From each of these divisions draw lines parallel to the base, as shown.

Divide the top and bottom lines of the right-hand square also into $\frac{1}{10}$ths of an inch.

Join these "diagonally" as shown in the diagram and label the scale with the figures indicated.

This scale will then measure accurately to $\frac{1}{100}$ in. (0·01 in.).

This will be readily understood if we consider the triangle Oab (Fig. *g*).

The distance from the 3 in. mark to the point O, along the bottom horizontal line, is obviously 3 in.

The distance from the 3 in. mark to the point b, along the top horizontal line, is 3·1 in., since $ab = 0{\cdot}1$ in.

We know that the ten horizontal lines are spaced at intervals of $\frac{1}{10}$ in., so that the length from the 3 in. mark along the first horizontal to the intersection with the diagonal Ob is $3 + (\frac{1}{10}$ of $0{\cdot}1)$ in. $= 3{\cdot}01$ in.

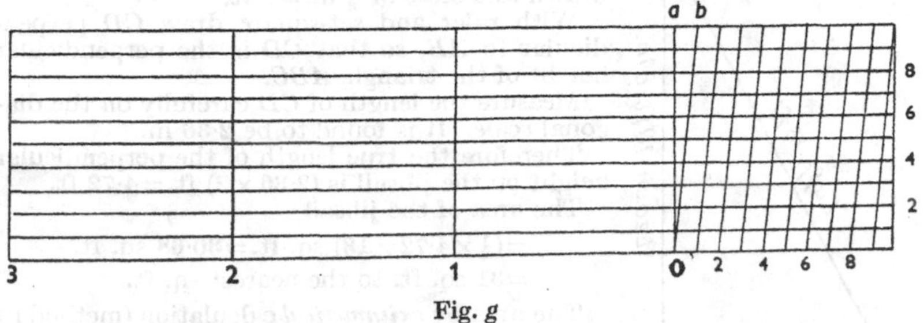

Fig. *g*

The distance from the 3 in. mark along the number 2 horizontal to its intersection with the diagonal Ob is $3 + (\frac{2}{10} \times 0{\cdot}1)$ in. $= 3{\cdot}02$ in., and so on.

To measure a length equal to 2·64 in. place the point of the dividers or compasses on the point of intersection of the number 6 "diagonal" with the number 4 horizontal. Open the compasses along the number 4 horizontal to the 2 in. mark. The distance thus measured is 2·64 in.

To measure a line of unknown length, first determine its approximate length with a ruler. If greater than 4 in. (since this is the maximum measurement we can make directly with the diagonal scale as drawn in Fig. *g*), mark off a whole number of inches, by the ruler, so that the remainder is less than 4 in.

Suppose, for example, that the line is more than 8 in. long, then mark off 5 in. by ruler, leaving between 3 and 4 in. to be measured.

Open the compasses accurately over this remaining length and with the point on the 3 in. mark of the diagonal scale determine which "intersection" agrees with the measured distance. If the agreement is along the number 7 horizontal at the intersection with the number 5 "diagonal", then the length measured is 3·57 in., and the length of the line in question is $5 + 3{\cdot}57 = 8{\cdot}57$ in.

Example. A jibsail is of the following dimensions: luff, 13 ft.; leach, 10 ft. 6 in.; foot, 6 ft. What is the sail area to the nearest square foot? (See Fig. *h*.)

Select, first, a suitable scale (say $\frac{1}{2}$ in. $= 1$ ft.), and draw the jibsail accurately to scale.

The luff, drawn to scale, is $\frac{13}{2}$ in. $= 6{\cdot}5$ in.

The leach, ,, $\dfrac{10\frac{1}{2}}{2} = \dfrac{21}{4} = 5{\cdot}25$ in.

The foot, ,, $\frac{6}{2} = 3$ in.

Draw a base line AB, 6·5 in. long, representing the length of the luff. Then with compasses open to 5·25 in. and centre at A make an arc. With compasses open to 3 in. and centre at B make another arc cutting the first arc at C (Fig. *i*).

The triangle ABC then represents the jibsail drawn to a scale of $\frac{1}{2}$ in. = 1 ft.

With ruler and set-square draw CD perpendicular to AB, so that CD is the perpendicular height of the triangle ABC.

Measure the length of CD carefully on the diagonal scale. It is found to be 2·36 in.

Therefore the true length of the perpendicular height on the jibsail is $(2{\cdot}36 \times 2)$ ft. $= 4{\cdot}72$ ft.

The area of the jibsail

$$= (\tfrac{1}{2} \times 4{\cdot}72 \times 13) \text{ sq. ft.} = 30{\cdot}68 \text{ sq. ft.}$$

$$= 31 \text{ sq. ft. to the nearest sq. ft.}$$

Fig. *h*

The area by *arithmetical* calculation (method ii) is obtained from a square-root formula.

It is more accurate than the scale-drawing method, but if the dimensions include many fractions it is somewhat laborious. The conventional lettering of a triangle is shown in Fig. *j*.

The angles are referred to by the capital letters, i.e. angle A, angle B and angle C.

The sides are referred to by small letters corresponding to the side opposite the angle with the same capital letter, i.e. side a opposite angle A, side b opposite angle B and side c opposite angle C.

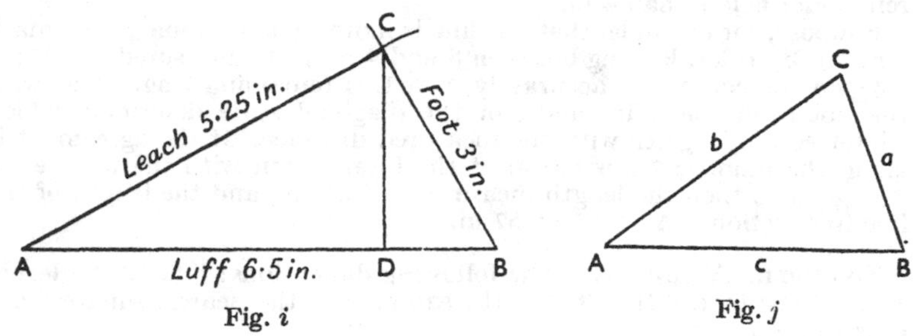

Fig. *i* Fig. *j*

The sum of the lengths of all three sides of a triangle (i.e. its perimeter) $= a + b + c$.

And half the perimeter $= \dfrac{a+b+c}{2}$.

If we represent this value, $\dfrac{a+b+c}{2}$, by the letter s, then the formula for finding the area of a triangle is

$$\text{Area} = \sqrt{s(s-a)(s-b)(s-c)}.$$

The perimeter of the jibsail (whose area we have just estimated by scale drawing) is

$$(13 + 10\tfrac{1}{2} + 6)\text{ ft.} = 29\tfrac{1}{2}\text{ ft.}$$

So that s (half the perimeter) $= 14\frac{3}{4}$ ft.
Therefore the area of the jibsail

$$= \sqrt{14\tfrac{3}{4}(14\tfrac{3}{4}-6)(14\tfrac{3}{4}-10\tfrac{1}{2})(14\tfrac{3}{4}-13)}$$

$$= \sqrt{14\tfrac{3}{4} \times 8\tfrac{3}{4} \times 4\tfrac{1}{4} \times 1\tfrac{3}{4}}$$

$$= \sqrt{\frac{59 \times 35 \times 17 \times 7}{256}}$$

$$= \sqrt{960}\text{ (nearly)} = 31\text{ sq. ft. to nearest sq. ft.}$$

A mainsail, or any irregular four-sided area, may be divided into triangles and drawn to scale or its area calculated from the formula, provided that the length of one diagonal is known.

The names of the various parts of a mainsail are given in Fig. k.

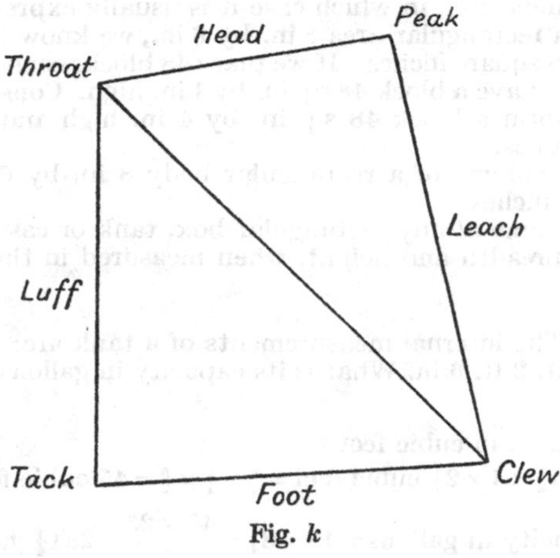

Fig. k

EXERCISE XLIX.

Find, by scale drawing, the sail area of all the following sails to the nearest square foot. Check the answers to questions 1, 2 and 5 by use of the above formula.

No.	Type of sail	Sail measurements				Diagonal Throat to clew
		Head	Foot	Luff	Leach	
		ft. in.	ft. in.	ft. in.	ft. in.	ft. in.
1	Jibsail	—	7 0	14 0	11 0	—
2	Jibsail	—	7 0	15 0	12 0	—
3	Jibsail	—	3 9	7 9	8 6	—
4	Trysail	—	11 6	12 0	15 8	—
5	Lugsail	7 0	8 0	5 6	11 6	10 0
6	Mainsail	15 8	12 2	4 9	12 6	10 4
7	Mainsail	17 10	15 2	4 8	14 3	13 8
8	Lugsail	9 0	11 6	8 6	16 6	14 6
9	Dipping lug	13 4	21 5	22 10	15 8	19 0
10	Dipping lug	13 6	12 3	10 2	21 10	15 7

Volume.

The volume of a body is the amount of space that it occupies, and is measured in cubic units of length, e.g. cubic inches, cubic feet, etc.

In the case of a hollow vessel, the term is used to define its capacity for storing liquids, etc., in which case it is usually expressed in gallons.

If we take a rectangular area 8 in. by 6 in., we know that this can be divided into 48 square inches. If we place 48 blocks each one inch square on this area, we have a block 48 sq. in. by 1 in. high. Consequently 4 such layers would form a block 48 sq. in. by 4 in. high and would contain (48×4) inch cubes.

So that the volume of a rectangular body 8 in. by 6 in. by 4 in. is $8 \times 6 \times 4$ cubic inches.

Thus the volume of any rectangular box, tank or case is the product of its length, breadth and height, when measured in the same units of length.

Example. The internal measurements of a tank are: base, 4 ft. 6 in. by 4 ft.; height, 2 ft. 6 in. What is its capacity in gallons, if 1 cubic foot $= 6\frac{1}{4}$ gallons?

Volume of tank in cubic feet

$$= 4\tfrac{1}{2} \times 4 \times 2\tfrac{1}{2} \text{ cubic feet} = \tfrac{9}{2} \times \tfrac{4}{1} \times \tfrac{5}{2} = 45 \text{ cubic feet.}$$

$$\therefore \quad \text{Capacity in gallons} = 45 \times 6\tfrac{1}{4} = \frac{45 \times 25}{4} = 281\tfrac{1}{4} \text{ gallons.}$$

Buoyancy.

A body is said to have a positive buoyancy if its weight is *less* than the weight of an equal volume of the liquid in which it is immersed. In other words, if a body has a positive buoyancy it will float.

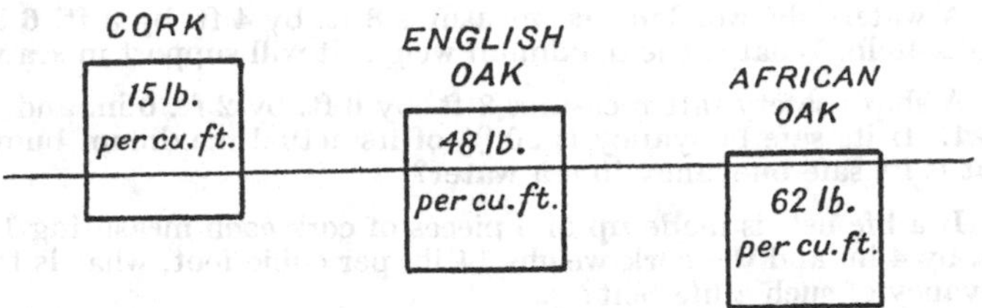

SEA WATER 64 lb. per cu. ft.

In the above figure are shown three foot cubes floating in sea water.

The cork cube displaces only $\frac{15}{64}$ of a cubic foot of sea water.

The English oak displaces $\frac{48}{64}$ and the African oak $\frac{62}{64}$ of a cubic foot.

Thus a cubic foot of cork can support a weight of $(64-15)=49$ lb. in sea water without sinking.

A cubic foot of English oak can support, under the same conditions, $(64-48)=16$ lb., and a cubic foot of African oak $(64-62)=2$ lb.

This extra weight that any floating substance can support in a liquid is known as its "reserve of buoyancy" in that liquid and is usually expressed as lb. per cubic foot.

In fresh water the reserve of buoyancy of cork would be

$$(62{\cdot}5-15)=47{\cdot}5 \text{ lb. per cubic foot.}$$

Naturally, a body that sinks in a liquid has a negative buoyancy in that liquid.

EXERCISE L.

1. A tank for water storage is required to hold 2000 gallons. If its height must be 5 ft., what must be the area of its base? (1 cu. ft. $= 6\frac{1}{4}$ gal.)

2. What is the height of a tank whose base is 7 ft. by 4 ft. if it just holds 525 gallons?

3. A vessel has a gross tonnage of 12,000 tons. The deduction allowed for crew space is 4 % of this. If the height of the crew's quarters between decks is 8 ft., what floor area is allocated to the crew?

4. A rectangular compartment in a ship measures 24 ft. by 10 ft. by 8 ft. Find (*a*) its volume in cubic feet, (*b*) the number of gallons of sea water it will hold when flooded, (*c*) the weight, in tons, of the sea water that floods it. (1 cu. ft. of sea water weighs 64 lb.)

5. One hold of a grain-loading vessel measures 44 ft. by 20 ft. by 10 ft. She is taking in oats which stow at 80 cu. ft. to the ton in bulk, but require 10 % extra space per ton when loaded in bags. Find the weight of oats in the hold when loaded (*a*) in bulk, (*b*) in bags.

6. A ship's hold can carry 160 tons of sugar in bags. The sugar stows at 50 cu. ft. per ton. If the depth of the hold is 8 ft. and its breadth is 20 ft., what is its length?

7. A watertight wooden case measures 8 ft. by 4 ft. by 3 ft. 6 in. and weighs 48 lb. What is the maximum weight it will support in sea water?

8. A ship's safety raft measures 8 ft. by 6 ft. by 2 ft. 6 in. and weighs 5 cwt. If its safe buoyancy is 90 % of its actual maximum buoyancy, what is its safe buoyancy in sea water?

9. If a life belt is made up of 6 pieces of cork each measuring 1 ft. by 6 in. by 4 in. and the cork weighs 14 lb. per cubic foot, what is the safe buoyancy of such a life belt?

10. Fill in the blank spaces in the following table. The buoyancy reserve to be stated as lb. per cubic foot in sea water. (Answers to nearest first decimal place.)

	Wood	lb./cu. ft.	cu. ft./ton	Buoyancy reserve
(*a*)	Ash	48		
(*b*)	Box		36·7	
(*c*)	Chestnut		59·0	
(*d*)	Deal	31		
(*e*)	Elm	45		
(*f*)	Ironwood		31·55	
(*g*)	Mahogany			+11
(*h*)	Lignum V.			−20
(*j*)	Teak		37·33	
(*k*)	Pine	30		

Burden Tonnage of Lighters and Barges.

This is a tonnage calculated from weight-carrying capacity, in a manner similar to the calculation of gross tonnage.

Suppose that the dimensions of a lighter are: length, L feet; breadth, B feet and depth, D feet. Then the volume of a rectangular box which would just enclose the lighter would be $L \times B \times D$ cubic feet.

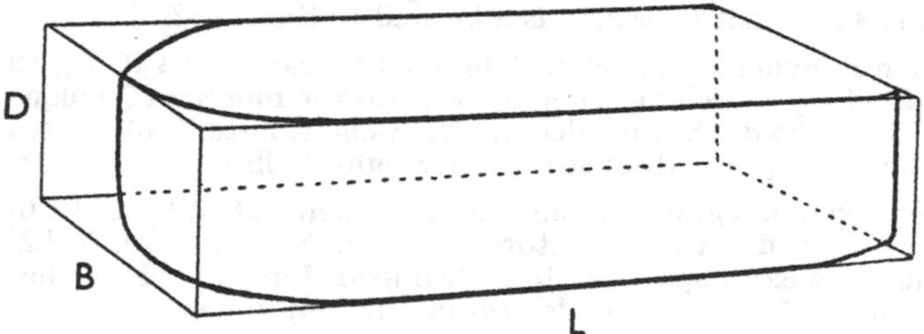

The volume of the lighter, by reason of its shape being not a true rectangular block, is less than LBD cubic feet.

Its approximate volume is determined by multiplying LBD cubic feet by a "coefficient of fineness", which is usually about 0·8. i.e.

Approximate volume of lighter $= 0{\cdot}8 \times LBD$ cubic feet.

So that the gross tonnage of the lighter $= \dfrac{0{\cdot}8 \times LBD}{100}$ tons.

The burden tonnage is generally reckoned to be $1\frac{2}{3}$ (i.e. 1·67) times the gross tonnage.

Thus the burden tonnage, or weight-carrying capacity, of a lighter measuring L feet by B feet by D feet, is $\dfrac{0{\cdot}8 \times 1{\cdot}67 \times LBD}{100}$ tons.

Exercise LI.

Fill in the missing details in the following table. Express all dimensions in feet to the first decimal place, and burden tonnage to the nearest ton.

No.	Length in ft.	Breadth in ft.	Depth in ft.	Burden tonnage
1	70	15	10	
2	85	10	6	
3	80	16	7·6	
4	95	19	8·3	
5	132	26·4	10·4	
6		20	10	400·8
7	120	24		307·9
8	115		9·9	365
9		21	11	478·4
10	123	20		300

Dimensions of Boats.

The Board of Trade standard measurements for boats are based upon the following formulae:

(1) For boats between 22 ft. and 24 ft. in length, the minimum breadth is 7 ft. 6 in.

(2) For boats of 24 ft. in length or over, the minimum breadth is $\dfrac{\text{Length in ft.} + 6\text{ ft.}}{4}$.

(3) For boats of 22 ft. in length or under, the minimum breadth is $\dfrac{\text{Length in ft.} + 7\text{ ft.}}{4}$.

(4) The maximum depth of any boat must not exceed

$$\frac{0{\cdot}42\,(\text{Length in ft.} + 6\text{ ft.})}{4}.$$

(5) The coefficient of fineness for any such boat is 0·6.

Exercise LII.

By reference to the above formulae, calculate the details necessary to complete the following table. Express dimensions in feet and inches to the nearest inch permissible, and capacity to the nearest cubic foot.

No.	Length in ft.	Minimum breadth	Maximum depth	Capacity in cubic feet
1	9			
2	26			
3	23			
4	18			
5	30			
6	16			
7	27			
8	40			
9	15			
10	32			

GRAPHS

STATISTICAL GRAPHS

Plotting of Points.

It is a common practice to demonstrate statistical details or recorded facts by means of "graphs" (or more correctly "charts"). A true graph is explained later.

For this purpose it is most convenient to use "squared" paper, ruled in tenths of an inch or in centimetres and millimetres.

Example. Suppose that we wish to record the noon to noon daily runs of a ship at sea, from 3 June to 11 June inclusive.

The details of these daily runs are given in the following table:

June	3	4	5	6	7	8	9	10	11
Distance run in N.M.	—	260	242	270	264	284	255	250	276

Draw upon the squared paper two lines at right angles, *OA* and *OB*, called axes.

The point of intersection *O* is called the point of origin, from which measurements are made.

The horizontal axis *OA* is used for recording the days, from 3 June to 11 June, and the vertical axis *OB* is used to record the distances run.

On the horizontal axis select a suitable scale, say five small squares, to represent one day, and on the vertical axis a scale, say five small squares, to represent 10 nautical miles, and label the points 220 to 300, as shown.

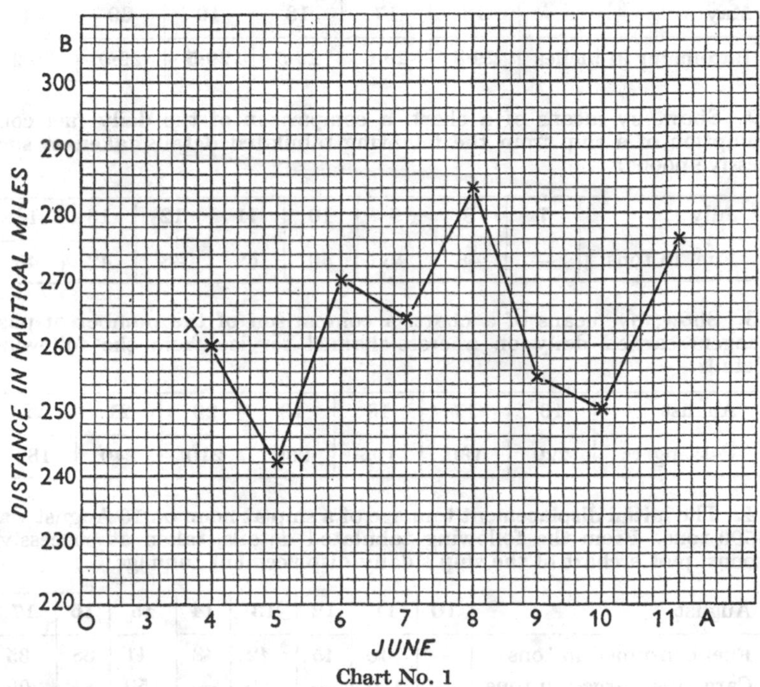

Chart No. 1

The first point of the chart (260 miles run up to noon 4 June) is obtained by travelling horizontally from the 260 mark on the vertical scale to the point immediately above the 4 June mark on the horizontal scale. This is the point *X* on the chart. The point *Y* is found in a similar manner from the 242 mark on the vertical scale to a point immediately above the 5 June mark—and so on. These points are usually joined together by straight lines which serve as a means to carry the eye from one point to the next.

It must be realised that such a chart as this has no value except as a record of fact. It is of no use for estimating or forecasting.

Although the distance run from noon on 4 June to noon on 5 June was 242 nautical miles, it does *not* follow that by midnight on 4/5 June the distance run was 121 nautical miles.

Exercise LIII.

1. Plot a barometric pressure chart from the following barometer readings taken at noon on each of the days shown:

May	16	17	18	19	20	21
Barometer in inches	29·8	29·6	29·5	29·5	29·9	30·2

2. Show, by means of a chart, a comparison of the daily fuel consumption of a ship from the following tabulated details taken at successive noons:

July	7	8	9	10	11	12	13	14
Fuel in tons	—	48	45	50	52	48	42	40

3. Show, by means of a chart, a comparison of the number of passengers carried daily on a cross-channel service from the following details:

August	16	17	18	19	20	21	22
Passengers	220	180	195	236	210	240	186

4. The initial displacement tonnage of a ship at noon on 10 August was 4500 tons. From the following tabulated details, taken at successive noons, plot a chart of the ship's daily displacement tonnage:

August	10	11	12	13	14	15	16	17
Fuel consumed in tons	—	40	45	42	48	41	38	35
Cargo discharged in tons	—	120	—	60	—	150	—	200
Cargo shipped in tons	—	—	—	220	—	80	—	—

CONTINUOUS OR LOCUS GRAPHS

Straight Line Graphs.

These are true graphs, and differ from the statistical charts so far dealt with in one important respect. They *are* used to determine intermediate values from known recorded facts.

When two quantities are known to be directly proportional to each other, then the graph showing the connection between these two quantities is a straight line which passes through the origin.

Example. A vessel is steaming at a steady 20 knots. Show, by means of a graph, the distance run at any interval of time up to 1 hr.

We know that in 6 min. the distance covered is $\frac{20}{10}=2$ nautical miles, and in 12 min. (twice the time) the distance run is $\frac{20}{5}=4$ nautical miles (twice the distance).

These quantities, therefore, are in direct proportion.

To plot this as a graph, first draw the two axes *OA* and *OB* as before, with the horizontal axis representing time intervals of one minute, and the vertical axis representing distance run in nautical miles (Graph No. 1).

The scales chosen are:

1 small square = 1 min. of time along the horizontal axis.

2 small squares = 1 N.M. along the vertical axis.

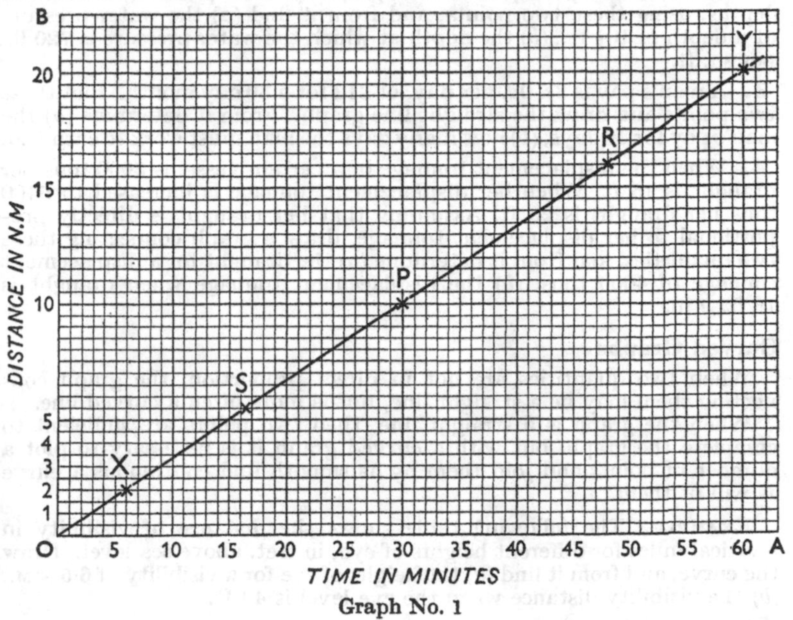

Graph No. 1

Plot the two points *X* (2 N.M. in 6 min.) and *Y* (20 N.M. in 60 min.).

Any other two points bearing the correct relationship would serve equally well.

Join the points *Y*, *X* by a straight line and note that this line when continued passes through the point of origin *O*, which illustrates that the distance covered in 0 min. is 0 N.M., which is, of course, true. Distances run in any interval of time up to 1 hr. may then be read directly from the graph.

For example, the point P shows that in 30 min. the distance run is 10 N.M.
The point R shows that in 48 min. the distance run is 16 N.M.
And the point S shows that in $16\frac{1}{2}$ min. the distance run is $5\frac{1}{2}$ N.M.

EXERCISE LIV.

1. Assuming that 1 cubic foot $= 6\frac{1}{4}$ gallons, draw a graph connecting cubic feet and gallons between the limits of 4 cu. ft. and 8 cu. ft. From the graph find (*a*) the number of gallons in $6\frac{1}{2}$ cu. ft., (*b*) the number of cubic feet that are equivalent to 60 gallons.

2. The pressure of sea water increases with depth by 64 lb. per sq. ft. for every foot increase in depth. Thus at 6 ft. below the surface the water pressure is 384 lb. per sq. ft., and at 15 ft. below the surface the pressure is 960 lb. per sq. ft. Plot a graph of water pressure at different depths, using these two points, and from it find (*a*) the water pressure at a depth of $10\frac{1}{2}$ ft., (*b*) the depth at which the water pressure is 720 lb. per sq. ft.

3. A man's pay is 8*s*. 0*d*. per day, or £12 for a 30 day month. Plot these two points and draw the straight line graph. From it determine (*a*) the man's pay for 40 days, (*b*) how many days the man must work to earn £20.

4. When the displacement tonnage of a certain vessel is 3000 tons her draught is 15 ft. When her displacement tonnage is increased to 5400 tons her draught is 20 ft. Assuming that her draught is directly proportional to her displacement tonnage, draw a graph connecting these two quantities, and from it determine (*a*) the draught for a displacement tonnage of 4000 tons, (*b*) the displacement tonnage when draught is 16 ft. 9 in.

Curved Graphs.

When two quantities are not in direct proportion, the graph connecting them may be a straight line, but is more often a curved line.
When the graph is a straight line, then two points are sufficient to complete the graph, but with a curved graph it is necessary to plot a number of points and join them by as smooth and continuous a curve as will fit them.

EXAMPLE. The following table shows the distance of visibility in nautical miles for different heights of eye, in feet, above sea level. Draw the curve, and from it find (*a*) the height of eye for a visibility of 6·5 N.M., (*b*) the visibility distance when the eye level is 43 ft.

Height of eye in feet	Visibility distance in N.M.
10	3·6
20	5·1
30	6·3
40	7·3
50	8·1
60	8·9

On the horizontal axis OA select a suitable scale, 1 small square $=0\cdot1$ N.M., to represent visibility distance in nautical miles.

On the vertical axis OB select a scale, 1 small square $=1$ ft., to represent height of eye in feet.

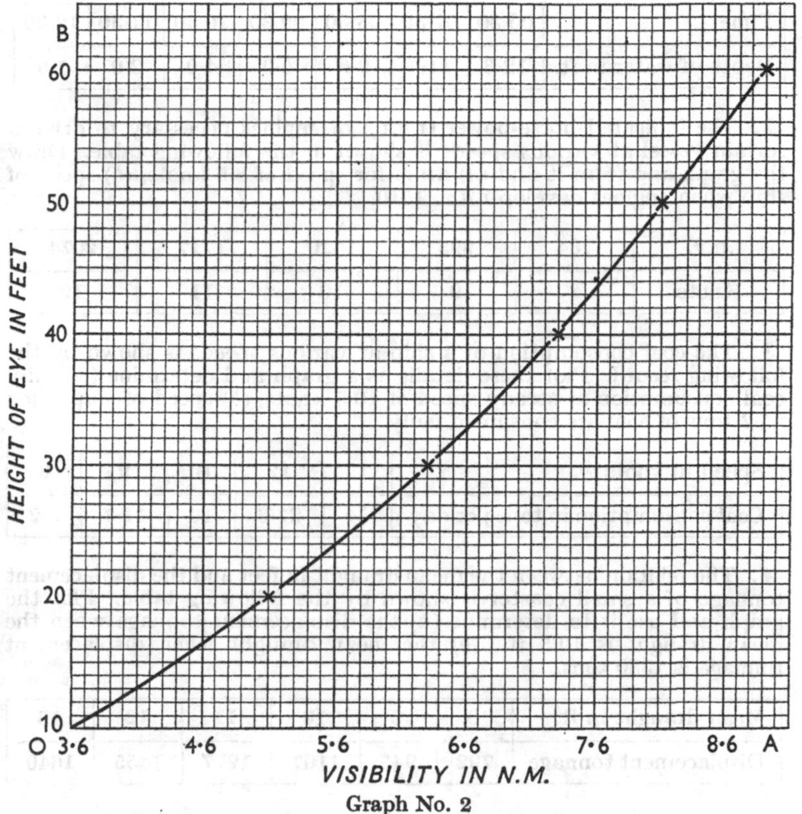

Graph No. 2

Plot the six separate points as given in the table and join these by a smooth curve, as shown.

It is convenient, in a case such as this, to label the point of origin on the horizontal axis as the first point (3·6) and also as the first point (10) on the vertical axis. This saves unnecessary waste of space when plotting the graph.

By reference to the graph the answers required are (*a*) 31·5 ft., (*b*) 7·55 N.M.

EXERCISE LV.

1. The following is a record of the depth of water, on 1 July, between 6.30 a.m. and 12.30 p.m. at Harwich. Plot the graph and find (*a*) depth at 07 42, (*b*) time when depth was 5·7 ft.

Time	06 30	07 30	08 30	09 30	10 30	11 30	12 30
Depth of water in ft.	12·3	11·5	9·6	6·9	4·0	2·0	1·3

2. The indicated horse-power (I.H.P.) of engines necessary to drive a certain vessel at various speeds is shown in the following table. Draw the graph and from it find (*a*) I.H.P. for speed of 8·5 knots, (*b*) speed of ship when engines develop I.H.P. 1510.

I.H.P.	734	995	1320	1717	2074
Knots	8	9	10	11	12

3. The coal consumption of a ship at various speeds is shown by the following record. Plot these details as a graph and determine (*a*) daily coal consumption at steady speed of 10·5 knots, (*b*) speed of ship when 35·2 tons of coal are consumed daily.

Speed in knots	15	12	10	9	8
Coal consumption in tons per day	54	27·6	16	11·7	8·2

4. The relation between the mean draught in feet and the displacement tonnage of a small coaster is shown by the following table. Plot the graph and use it to determine (*a*) the displacement tonnage when the mean draught is 10·5 ft., (*b*) the mean draught when displacement tonnage is 1330 tons.

Mean draught in ft.	8	9	10	11	12	13
Displacement tonnage	792	945	1107	1277	1455	1640

NAVAL EDUCATIONAL TESTS FOR REVISION

The following educational tests are included through the courtesy of the Admiralty and the Controller of H.M. Stationery Office.

The instructions given with these papers are the same in all cases and are these:

Time allowed, $2\frac{1}{2}$ hours.

Write on both sides of the paper. Show all your rough work in the margin of the page. Unless you make plain the steps by which your answers are obtained, full marks cannot be earned. Answer not more than EIGHT *questions. The questions may be taken in any order, but must be numbered as below* (A1, A2, B3, B4, *etc.*).

JULY, 1940

A1. (*a*) Simplify $1\frac{1}{8}$ of $\frac{4}{27} \div 3\frac{3}{4}$ of $3\frac{1}{5}$.

(*b*) Four bells begin to toll simultaneously and toll at intervals of 16, 18, 20 and 25 seconds respectively. After what time will they again toll simultaneously?

A2. (*a*) Reduce $7\frac{1}{2}d.$ to the decimal of $15s.$ (Correct to three places.)

(*b*) Simplify $\dfrac{0{\cdot}003\,(4{\cdot}047-0{\cdot}001)}{0{\cdot}07}$.

A3. Twenty men started to dig a trench which would have taken them 10 hours to complete. When only one-fifth had been dug, five men were taken off. How much longer would it take the remaining fifteen men to finish the trench?

A4. A gift amounting to £928 was shared out at the following agreed ratios: Petty Officer 10, Leading Seaman 7, Able Seaman 5, Boy 3. If there were 3 P.O.s, 4 Ldg. Seamen, 30 A.B.s and 8 Boys, how much did each receive?

A5. A rectangular courtyard, 16 yards by $14\frac{1}{2}$ yards, has two small square ponds in it, each of $2\frac{1}{2}$ yards sides. Find the cost of cementing the courtyard at $6d.$ per square foot. (The ponds are not being emptied.)

A6. Which is the greater distance and by how many yards? Bristol to London by air (110 miles) or Havre to Paris by air (180 kilometres) if

1 kilometre = 1000 metres, 1 metre = 1·093 yards.

A7. A dealer made $2\frac{1}{2}d.$ profit by selling an article at $1s.$ How much per cent profit was this?

A8. During a certain month a messman purchased the following items: 16 lb. butter at 1*s*. 7*d*. per lb.; 27 lb. margarine at $7\frac{1}{2}d$. per lb; 14 doz. eggs at $2\frac{1}{2}d$. each; 50 lb. fish at 8*d*. per lb.; 16 rabbits at 1*s*. 4*d*. each; 28 gallons of milk (in 1 pint bottles) at 2*s*. 4*d*. per gallon.

Find the total cost after allowing for the following rebates:

(1) Discount at 6*d*. in the £ for each complete £ in the above list.
(2) $\frac{1}{4}d$. for each glass milk bottle (none broken).
(3) Half-a-crown for returned crates.

A9. A mess of 70 men decided to give a Children's Party. Each member paid in 4*s*. and the mess voted £12 from funds.

Find debt or credit carried forward to mess funds after meeting the following expenses: Transport £5, 25 books at 9*d*. each, 18 games at $7\frac{1}{2}d$. each, 36 toys at $6\frac{1}{2}d$. each, catering £16. 15*s*. 9*d*., music £1. 10*s*. 0*d*., 30 vanity sets for mothers at 1*s*. 3*d*. each, 79 surprise bags of sweets, etc. at $5\frac{1}{2}d$. each.

July, 1940

B1. (*a*) Simplify $4\frac{1}{2}$ of $\frac{7}{36} \div 2\frac{1}{5}$ of $3\frac{3}{4}$.

(*b*) Find the least sum of money which contains 14*s*., 36*s*. and 21*s*. each an exact number of times.

B2. (*a*) Reduce $8\frac{1}{4}d$. to the decimal of 17*s*. 6*d*. (Correct to three places.)

(*b*) Simplify $\dfrac{0{\cdot}02\,(1{\cdot}086 - 0{\cdot}0005)}{0{\cdot}005}$.

B3. Sixteen men started to dig a trench which would have taken them 8 hours to complete. When only a quarter had been dug, four men were taken off. How much longer would it take the remaining twelve men to finish the trench?

B4. A gift amounting to £640 was shared out at the following agreed ratios: Petty Officer 8, Leading Seaman 6, Able Seaman 4, Boy 2. If there were 2 P.O.s, 4 Ldg. Seamen, 24 A.B.s and 12 Boys, how much did each receive?

B5. A rectangular courtyard, 18 yards by $15\frac{1}{2}$ yards, has two small square ponds in it, each of $3\frac{1}{2}$ yards sides. Find the cost of cementing the courtyard at 6*d*. per square foot. (The ponds are not being emptied.)

B6. Which is the greater distance and by how many yards? Exeter to London by air (150 miles) or Calais to Paris by air (240 kilometres) if

$$1 \text{ kilometre} = 1000 \text{ metres}, \quad 1 \text{ metre} = 1{\cdot}093 \text{ yards}.$$

B7. A dealer made $4\frac{1}{2}d$. profit by selling an article at 1*s*. 6*d*. How much per cent profit was this?

B8. During a certain month a messman purchased the following items: 15 lb. butter at 1*s.* 7*d.* per lb.; 25 lb. margarine at $7\frac{1}{2}$*d.* per lb.; 15 doz. eggs at $2\frac{1}{2}$*d.* each; 60 lb. fish at 8*d.* per lb.; 15 rabbits at 1*s.* 3*d.* each; 26 gallons of milk (in 1 pint bottles) at 2*s.* 4*d.* per gallon.

Find the total cost after allowing for the following rebates:

(1) Discount at 6*d.* in the £ for each complete £ in the above list.
(2) $\frac{1}{2}$*d.* for each glass milk bottle (none broken).
(3) Half-a-crown for returned crates.

B9. A mess of 60 men decided to give a Children's Party. Each member paid in 4*s.* and the mess voted £10 from funds.

Find debt or credit carried forward to mess funds after meeting the following expenses: Transport £3, 15 books at 9*d.* each, 16 games at $7\frac{1}{2}$*d.* each, 24 toys at $6\frac{1}{2}$*d.* each. Catering £11. 17*s.* 6*d.*, music, one guinea, 27 vanity sets for mothers at 1*s.* 3*d.* each, 55 surprise bags of sweets, etc. at $5\frac{1}{2}$*d.* each.

November, 1940

A1. (*a*) Simplify $(3\frac{1}{4}+2\frac{1}{16})\div(3\frac{1}{4}-2\frac{1}{16})$.

(*b*) A man left $\frac{2}{5}$ of his estate to his son, and $\frac{1}{3}$ of the remainder to his daughter. If the daughter received £400, find the total value of the estate.

A2. (*a*) Simplify $\dfrac{1{\cdot}37\times 4{\cdot}068}{5{\cdot}48}$.

(*b*) Find the value of £0·65 + 8·125*s.* − 10·25*d.*

A3. The number of men accommodated in a mess for the seven days of a week was 36, 34, 38, 27, 30, 28, 31. If each man consumed 36 ounces of food per day, find in pounds the average weight of food consumed per day.

A4. A ship takes 35 hours to reach a rendezvous steaming at 20 knots. At what speed must she steam to complete the journey in 25 hours? How long does she take if she steams at 30 knots?

A5. (*a*) A motor car uses 24 gallons of petrol in travelling a distance of 720 miles. How many gallons will it require for a journey of 1000 miles?

(*b*) A ship's company of 1200 men has provisions sufficient to last 12 weeks. How many men would eat this supply in 8 weeks?

A6. A man arranged in his will for his money to be divided so that his wife should receive 10 shares, each of his four sons 6 shares, and each of his three daughters 4 shares. If the total sum to be divided amounted to £18,400, how much did the wife, each son and each daughter receive?

A7. (*a*) Express as percentages the fractions $\frac{1}{8}$, $\frac{3}{4}$, 0·375.

(*b*) A builder charges £84 for building a shed. Of this sum, materials cost £24. 5*s*. 0*d*. and labour £38. 15*s*. 0*d*. Estimate his percentage profit.

A8. During a month a messman took up stores which exceeded his allowance due from the Paymaster by £4. 12*s*. 6*d*., and in addition made the following outside purchases: $1\frac{1}{2}$ cwt. new potatoes at $2\frac{1}{4}d$. per lb.; 50 lb. fresh haddock at 8*d*. per lb.; 26 lb. plaice at 1*s*. 1*d*. per lb.; 65 lb. new peas at $2\frac{1}{2}d$. per lb.; 70 lb. apples at $3\frac{1}{4}d$. per lb.; 4 doz. oranges at 1*s*. $2\frac{1}{2}d$. per doz.

Find his total costs for the month to the nearest penny, after deducting a discount of 5 per cent which he is allowed on outside purchases.

A9. A Petty Officer receives 6*s*. 5*d*. per day as pay, 2*s*. 0*d*. per day as Gunner's Mate, and 6*d*. per day for two G/C badges. He has been credited with 12 days' leave allowance at 2*s*. 2*d*. per day, and has been paid £18 during the quarter. His kit upkeep allowance for the quarter amounts to £1. 18*s*. 0*d*. and his quarterly allotment to £21. If his Insurance Contribution is 4*s*. 10*d*., find the amount due to him for the quarter ending 30 September.

NOVEMBER, 1940

B1. (*a*) Simplify $(3\frac{2}{5}+2\frac{1}{25})\div(3\frac{2}{5}-2\frac{1}{25})$.

(*b*) A man left $\frac{3}{7}$ of his estate to his son, and $\frac{1}{4}$ of the remainder to his daughter. If the daughter received £300, find the total value of the estate.

B2. (*a*) Simplify $\dfrac{1{\cdot}56\times 5{\cdot}024}{6{\cdot}24}$.

(*b*) Find the value of £0·95 + 7·375*s*. + 9·5*d*.

B3. The number of men accommodated in a mess for seven days of a week was 37, 33, 36, 29, 32, 27, 30. If each man consumed 36 ounces of food per day, find in pounds the average weight of food consumed per day.

B4. A ship takes 36 hours to reach a rendezvous steaming at 18 knots. At what speed must she steam to complete the journey in 24 hours? How long does she take if she steams at 21 knots?

B5. (*a*) A motor car uses 30 gallons of petrol in travelling a distance of 960 miles. How many gallons will it require for a journey of 640 miles?

(*b*) A ship's company of 850 men has provisions sufficient to last 15 days. How many men would eat this supply in 17 days?

B6. A man arranged in his will for his money to be divided so that his wife should receive 12 shares, each of his three sons 10 shares, and each of his four daughters 6 shares. If the total sum to be divided amounted to £13,200, how much did the wife ,each son and each daughter receive?

B7. (*a*) Express as percentages the fractions $\frac{1}{4}$, $\frac{17}{20}$, 0·625.

(*b*) A builder charges £114 for building a shed. Of this sum, materials cost £30. 16*s*. 0*d*. and labour £64. 4*s*. 0*d*. Estimate his percentage profit.

B8. During a month a messman took up stores which exceeded his allowance due from the Paymaster by £3. 15*s*. 2*d*., and in addition made the following outside purchases: $1\frac{1}{4}$ cwt. new potatoes at $2\frac{1}{4}$*d*. per lb.; 46 lb. fresh haddock at 8*d*. per lb.; 32 lb. plaice at 1*s*. 1*d*. per lb.; 56 lb. new peas at $2\frac{1}{2}$*d*. per lb.; 76 lb. apples at $3\frac{1}{4}$*d*. per lb.; 6 doz. oranges at 1*s*. $2\frac{1}{2}$*d*. per doz.

Find his total costs for the month to the nearest penny, after deducting a discount of 5 per cent which he is allowed on outside purchases.

B9. A Petty Officer receives 6*s*. 10*d*. per day as pay, 2*s*. 0*d*. per day as Gunner's Mate, and 6*d*. per day for two G/C badges. He has been credited with 10 days' leave allowance at 2*s*. 2*d*. per day, and he has been paid £18 during the quarter. His kit upkeep allowance for the quarter amounts to £1. 15*s*. 0*d*. and his quarterly allotment to £22. If his Insurance Contribution is 4*s*. 10*d*., find the amount due to him for the quarter ending 30 September.

March, 1941

A1. (*a*) Simplify $\dfrac{(2\frac{3}{4}+\frac{1}{4}) \text{ of } 2\frac{5}{8}}{1\frac{1}{4} \text{ of } (\frac{3}{2}+2\frac{1}{3})}$.

(*b*) Express £3. 6*s*. 8*d*. as a fraction of £5.

A2. (*a*) Simplify $\dfrac{4{\cdot}44 \times 0{\cdot}031 \times 0{\cdot}0001}{13{\cdot}024 + 0{\cdot}74}$.

(*b*) Reduce £3. 9*s*. 4*d*. to the decimal of £1. 7*s*. 1*d*.

A3. If 60 men could assemble ten Anderson shelters in 3 hours 5 minutes, how many men would be required to assemble fifteen shelters in 1 hour 51 minutes?

A4. (*a*) Arrange the following in order of magnitude, the largest first:

0·017, $\frac{3}{16}$, 2 per cent.

(*b*) A grocer blends 75 lb. of Ceylon tea with 59 lb. of China tea. What is the percentage of each in the mixture?

A5. Of a pole $\frac{1}{12}$ is painted white, $\frac{7}{30}$ green, $\frac{13}{60}$ red, and the remaining 5 feet black. Find the length of the pole.

A6. The average weight of the eight oarsmen in a boat is increased by 3 lb. when one of the crew who weighs 11 stone is replaced by a fresh man. What is the weight of the new man?

A7. A picture frame, 3 inches wide, surrounded a picture which measured 3 feet by 2 feet inside the frame. Find the cost of re-gilding the front of the framing at 2*d.* per square inch.

A8. A messman had to pay the Paymaster £3. 7*s.* 6*d.* for excess stores as well as meet the following items purchased ashore: 2 doz. pkts. cornflakes at 5*d.* per pkt.; 5 doz. apples at $1\frac{1}{2}d.$ each; 3 doz. oranges at 4*d.* each; 24 lb. fresh fish at 9*d.* per lb.; 2 crates of assorted fresh vegetables at 5*s.* 7*d.* per crate; 14 gallons milk at $3\frac{1}{2}d.$ per pint.
He was allowed the following rebates on his shore purchases:

(1) 6*d.* in the £, to the nearest £, discount.
(2) Five shillings for returned empties.

Find his total costs after settling all accounts.

A9. From the following particulars, estimate what the March 1941 quarterly settlement should be:

Pay, 6*s.* 6*d.* a day. Two badges at 3*d.* a day each. 1*s.* 6*d.* a day as T.G.M. Allowance in lieu of spirit ration, 3*d.* a day. Kit upkeep, £1. 15*s.* 0*d.* per quarter. Ten days' leave allowance at 2*s.* 2*d.* per day. Two advances during the quarter each of £6 and a monthly allotment to home of £8. Health Insurance of 4*s.* 10*d.* is to be deducted.

March, 1941

B1. (*a*) Simplify $\dfrac{(3\frac{1}{4}+\frac{3}{4}) \text{ of } 3\frac{1}{8}}{2\frac{1}{2} \text{ of } (\frac{4}{3}+3\frac{1}{6})}$.

(*b*) Express £2. 3*s.* 4*d.* as a fraction of £4.

B2. (*a*) Simplify $\dfrac{1000 \times 0{\cdot}0006 \times 9}{0{\cdot}22007 + 0{\cdot}04993}$.

(*b*) Reduce 15*s.* $11\frac{1}{4}d.$ to the decimal of 3*s.* 9*d.*

B3. If 30 men could assemble ten Anderson shelters in 3 hours 42 minutes, how many men would be required to assemble twelve shelters in 2 hours 28 minutes?

B4. (*a*) Arrange the following in order of magnitude, the largest first:

$0{\cdot}024$, $\frac{2}{25}$, 1 per cent.

(*b*) A grocer makes a blend containing 73 lb. of Ceylon tea and 56 lb. of China tea. What is the percentage of each in the mixture?

B5. Of a farm $\frac{4}{15}$ is planted with turnips, $\frac{1}{4}$ with potatoes, $\frac{7}{60}$ with beet, and the remaining 4 acres left for grazing. Find the size of the farm in acres.

B6. The average weight of the eight oarsmen in a boat is increased by 2 lb. when one of the crew who weighs 12 stone is replaced by a fresh man. What is the weight of the new man?

B7. A picture frame, 2 inches wide, surrounded a picture which measured 3 feet by 3 feet inside the frame. Find the cost of re-gilding the front of the framing at 2*d.* per square inch.

B8. A messman had to pay the Paymaster £4. 9*s.* 4*d.* for excess stores as well as meet the following items purchased ashore: 3 doz. pkts. cornflakes at 5*d.* per pkt.; 4 doz. apples at $1\frac{1}{2}$*d.* each; 4 doz. oranges at 4*d.* each; 20 lb. fresh fish at 9*d.* per lb.; 2 crates of assorted vegetables at 5*s.* 9*d.* per crate; 12 gallons milk at $3\frac{1}{2}$*d.* per pint.
He was allowed the following rebates on his shore purchases:

(1) 6*d.* in the £, to the nearest £, discount.
(2) Five shillings for returned empties.

Find his total costs after settling all accounts.

B9. From the following particulars, estimate what the March 1941 quarterly settlement should be:

Pay, 6*s.* 8*d.* a day. Three badges at 3*d.* a day each. 1*s.* 6*d.* a day as T.G.M. Allowance in lieu of spirit ration, 3*d.* a day. Kit upkeep, £1. 15*s.* 0*d.* per quarter. Twelve days' leave allowance at 2*s.* 2*d.* a day. Two advances during the quarter each of £5. Monthly allotment to home, £9. Health Insurance of 4*s.* 10*d.* is to be deducted.

JULY, 1941

A1. (*a*) Simplify $\dfrac{1\frac{1}{3}\times 1\frac{4}{5}\times 1\frac{7}{18}}{3\frac{1}{3}-1\frac{1}{9}}$.

(*b*) 18 ft. of a post is above ground, and $\frac{1}{3}$ of it is buried underground. Find the total length of the post.

A2. (*a*) Find the value of $3{\cdot}185-2{\cdot}34+1{\cdot}18-0{\cdot}696$.

(*b*) Express $\frac{17}{40}$ and $\frac{11}{25}$ as decimal fractions, and state which is the greater.

A3. A motor car travels at 48 miles per hour for 20 minutes, stops for 5 minutes, and then travels at 32 miles per hour for 15 minutes. Find the average speed for the whole time. How much farther would the car have travelled if it had continued at 48 miles per hour for the whole time?

A4. Prize money to the value of £616 was divided among four men, *A*, *B*, *C* and *D*, in the proportion of 12:9:5:2. Calculate what each man received.

A5. A motor car can travel 32 miles on 1 gallon of petrol. Find the cost of the petrol for a journey of 800 miles. One gallon of petrol costs 2*s*. 1*d*.

A6. If 1 metre = 39·37 inches, calculate to two decimal places how many centimetres are equal to one inch and hence find the area in square centimetres of a rectangular board which is 12 inches long and 8 inches wide. (1 metre = 100 centimetres.)

A7. (*a*) Express as percentages the following fractions:

$$\tfrac{1}{3},\ \tfrac{18}{25},\ 0{\cdot}325.$$

(*b*) A man earning £25 per calendar month has his salary increased by 15 per cent. How much extra will he earn in one year?

A8. During a month a messman took up stores which exceeded his allowance from the Paymaster by £3. 12*s*. 6*d*., and in addition made the following outside purchases: 48 lb. of apples at $3\tfrac{1}{4}d$. per lb.; 6 doz. oranges at 1*s*. 8*d*. per doz.; 80 lb. of fish at a flat rate of $11\tfrac{1}{4}d$. per lb.; $2\tfrac{1}{2}$ cwt. of potatoes at 10*d*. for 14 lb.; 25 rabbits at 1*s*. 6*d*. each; 35 cauliflowers at $3\tfrac{1}{2}d$. each.

Find his total costs for the month. If there were 25 men in the mess, how much would each have to pay?

A9. Mess Account for July. 22 men in the Mess.

	£	*s*.	*d*.
Messing allowance	25	10	0
Special dinner vouchers		13	2
Provisions issued on repayment	13	6	5
Mess traps lost		10	4
Canteen bill	11	1	3
Extra expenses during month	2	18	3
Due from messmates for supplies		16	4

Find the amount each man must pay to keep the mess out of debt.

July, 1941

B1. (*a*) Simplify $\dfrac{2\frac{1}{3}\times\frac{5}{14}\times 3\frac{1}{5}}{3\frac{1}{3}-2\frac{2}{9}}$.

(*b*) 40 ft. of a post is above ground, and $\tfrac{1}{5}$ of it is buried underground. Find the total length of the post.

B2. (*a*) Find the value of $3{\cdot}225-1{\cdot}765+2{\cdot}432-2{\cdot}75$.

(*b*) Express $\tfrac{33}{80}$ and $\tfrac{12}{25}$ as decimal fractions, and state which is the greater.

B3. A motor car travels at 48 miles per hour for 25 minutes, stops for 4 minutes, and then travels at 60 miles per hour for 16 minutes. Find the average speed for the whole time. How much less distance would the car have travelled if it had continued at 45 miles per hour for the whole time?

B4. Prize money to the value of £550 was divided among four men, *A*, *B*, *C* and *D*, in the proportion of 10:8:5:2. Calculate what each man received.

B5. A motor car can travel 28 miles on 1 gallon of petrol. Find the cost of the petrol for a journey of 784 miles. One gallon of petrol costs 2*s.* 1*d.*

B6. If 1 metre = 39·37 inches, calculate to two decimal places how many centimetres are equal to one inch, and hence find the area in square centimetres of a rectangular board which is 14 inches long and 10 inches wide. (1 metre = 100 centimetres.)

B7. (*a*) Express as percentages the following fractions:

$\frac{1}{8}$, $\frac{25}{30}$, 0·665.

(*b*) A man earning £24 per calendar month has his salary increased by $33\frac{1}{3}$ per cent. How much extra will he receive in one year?

B8. During a month a messman took up stores which exceeded his allowance from the Paymaster by £2. 10*s.* 8*d.*, and in addition made the following outside purchases: 54 lb. of apples at $3\frac{1}{2}d.$ per lb.; 7 doz. oranges at 1*s.* 7*d.* per doz.; 64 lb. of fish at a flat rate of $11\frac{3}{4}d.$ per lb.; $1\frac{3}{4}$ cwt. of potatoes at 11*d.* for 14 lb.; 26 rabbits at 1*s.* 5*d.* each; 30 cauliflowers at $3\frac{1}{2}d.$ each.

Find his total costs for the month. If there were 24 men in the mess, how much would each have to pay?

B9. Mess Account for July. 24 men in the Mess.

	£	*s.*	*d.*
Messing Allowance	26	8	0
Special dinner vouchers		12	2
Provisions issued on repayment	14	10	8
Mess traps lost		12	8
Canteen bill	12	2	4
Extra expenses during month	2	19	6
Due from messmates for supplies		15	3

Find the amount each man must pay to keep the mess out of debt.

NOVEMBER, 1941

A1. (*a*) Simplify $\dfrac{3\frac{1}{4}-(1\frac{2}{3}+\frac{5}{6})}{1\frac{7}{8} \text{ of } 2\frac{1}{2}}$.

(*b*) A sack of coal originally contained 1 cwt. but, owing to a hole in it, when delivered it only contained 3 qr. 21 lb. Find what fraction of the coal had been lost in transit.

A2. (*a*) Simplify $\dfrac{(0{\cdot}005-0{\cdot}0005)+10{\cdot}5}{18{\cdot}75 \times 0{\cdot}008}$.

(*b*) Express 4 lb. 6 oz. as a decimal of 1 stone.

A3. Two hundred labourers were required to level a piece of land for an aerodrome in order to complete it in six months. After working for one month it was decided to finish the job during the next two months. How many more men would have to be signed on to do this?

A4. (*a*) An article was retailed for 6*s*. 8*d*. This price included "purchase tax" of $33\frac{1}{3}$ per cent. What was its price before taxation?

(*b*) The ordinary price of some goods was 18*s*. 4*d*., but the retailer had to add purchase tax at the rate of 15 per cent. What did the customer have to pay for the goods?

A5. (*a*) Express as fractions in their simplest terms:

0·02, 0·00005, 10 per cent, $12\frac{1}{2}$ per cent.

(*b*) Express as decimals:

$\frac{1}{16}$, $\frac{3}{5000}$, 1 per cent, $33\frac{3}{4}$ per cent.

A6. An estate was divided as follows after payment of all legal expenses, etc.: 0·025 to an orphanage; 0·04 to a Service Institute; 0·2 to an employees' benevolent fund. The rest had to be divided equally between three surviving children. If each of the children received £490, what total sum was distributed and what amount went to the Service Institute?

A7. In an international walking contest a competitor in England walked 160 yards in one minute, 150 in the next, 140 in the third and 135 in the fourth. A competitor in Egypt walked 150 metres in one minute, 140 in the next, 135 in the third and 130 in the fourth. Find the average distance per minute in each case, the first in yards per minute, the other in metres per minute. Also find which was the faster if 1 metre = 39·37 inches.

A8. From the following particulars, estimate a man's December quarterly settlement: Pay, 3*s*. 3*d*. a day. One badge, 3*d*. a day. Allowance in lieu of spirit ration, 3*d*. a day. Kit upkeep, 3*d*. a day. 6*d*. a day war bonus. Ten days' leave allowance at 3*s*. a day. He had received six fortnightly payments of 35*s*. each. He had made an allotment to home of 9*s*. per week. Health Insurance, 4*s*. 10*d*., is to be deducted.

A9. Compile a statement of accounts for a mess of twenty men from the following particulars and strike a balance showing how much each man must pay in order to clear all debt.

To Paymaster for stores issued in excess of allowance, £5. 12*s.* 6*d.* To a tradesman ashore, 14 lb. smoked ham at 2*s.* 3*d.* per lb., 16 tins assorted soups at 7½*d.* per tin, 9 lb. cheese at 1*s.* 2*d.* per lb., 6 lb. jam at 10½*d.* per lb., 20 lb. fresh fish at 1*s.* 2*d.* per lb., cleansing powders and soaps 5*s.* 9*d.*, miscellaneous goods £2. 3*s.* 4*d.* Deduct 6*d.* in the £ to the nearest £ for discount on the total of the tradesman's bill.

November, 1941

B1. (*a*) Simplify $\dfrac{4\frac{1}{4}-(2\frac{1}{3}+1\frac{1}{6})}{1\frac{5}{9} \text{ of } 2\frac{1}{4}}$.

(*b*) A sack of coal originally contained 1 cwt. but, owing to a hole in it, when delivered it only contained 3 qr. 18 lb. Find what fraction of the coal had been lost in transit.

B2. (*a*) Simplify $\dfrac{(0{\cdot}004-0{\cdot}0004)+10{\cdot}4}{1{\cdot}6\times 0{\cdot}025}$.

(*b*) Express 6 lb. 2 oz. as a decimal of 1 stone.

B3. Three hundred labourers were required to level a piece of land for an aerodrome in order to complete it in six months. After working for one month it was decided to finish the job during the next three months. How many more men would have to be signed on to do this?

B4. (*a*) An article was retailed for 9*s.* 4*d.* This price included "purchase tax" of 33⅓ per cent. What was its price before taxation?

(*b*) The ordinary price of some goods was 11*s.* 8*d.*, but the retailer had to add purchase tax at the rate of 15 per cent. What did the customer have to pay for the goods?

B5. (*a*) Express as fractions in their simplest terms:

0·04, 0·00008, 17½ per cent, 80 per cent.

(*b*) Express as decimals:

$\frac{1}{32}$, $\frac{7}{5000}$, 2 per cent, 55¼ per cent.

B6. An estate was divided as follows after payment of all legal expenses, etc.: 0·034 to an orphanage; 0·06 to a Service Institute; 0·3 to an employees' benevolent fund. The rest had to be divided equally between three surviving children. If each of the children received £606, what total sum was distributed and what amount went to the Service Institute?

B7. In an international walking contest a competitor in England walked 165 yards in one minute, 145 in the next, 150 in the third and 140 in the fourth. A competitor in Egypt walked 160 metres in one minute, 150 in the next, 140 in the third and 133 in the fourth. Find the average distance per minute in each case, the first in yards per minute, the other in metres per minute. Also find which was the faster if 1 metre = 39·37 inches.

B8. From the following particulars, estimate a man's December quarterly settlement: Pay, 3*s.* 6*d.* a day. Two badges at 3*d.* a day each. Allowance in lieu of spirit ration, 3*d.* a day. Kit upkeep, 3*d.* a day. 6*d.* a day war bonus. Twelve days' leave allowance at 3*s.* a day. He had received six fortnightly payments of 35*s.* each. He had made an allotment to home of 10*s.* per week. Health Insurance, 4*s.* 10*d.*, is to be deducted.

B9. Compile a statement of accounts for a mess of 24 men from the following particulars and strike a balance showing how much each man must pay in order to clear all debt.

To Paymaster for stores issued in excess of allowance, £6. 11*s.* 6*d.* To a tradesman ashore, 16 lb. smoked ham at 2*s.* 3*d.* per lb., 18 tins assorted soups at $7\frac{1}{2}$*d.* per tin, 11 lb. cheese at 1*s.* 2*d.* per lb., 8 lb. jam at $10\frac{1}{2}$*d.* per lb., 36 lb. fresh fish at 1*s.* 3*d.* per lb., cleansing powders and soaps 5*s.* 9*d.*, miscellaneous goods £2. 3*s.* 4*d.* Deduct 6*d.* in the £ to the nearest £ for discount on the total of the tradesman's bill.

March, 1942

A1. (*a*) Simplify $\dfrac{3\frac{5}{6}+2\frac{11}{30}}{3\frac{5}{6}-2\frac{11}{30}}$.

(*b*) If $\frac{3}{5}$ of a ship's cargo is worth £7860, calculate the value of the whole cargo.

A2. (*a*) Simplify $\dfrac{6{\cdot}81\times1{\cdot}145}{3{\cdot}45}$. Answer to be given to two decimal places only.

(*b*) Express £4. 12*s.* 6*d.* as a decimal fraction of £12. 6*s.* 8*d.*

A3. A ship steams at 24 knots for $8\frac{1}{2}$ hours, 30 knots for 2 hours, and 16 knots for $3\frac{3}{4}$ hours. How far did she travel, and what was the average speed for the whole time?

A4. What is the area of the walls of a room which is 18 feet long, 15 feet wide, and 12 feet high, and what would be the cost of painting the walls at 9*d.* per sq. yard?

A5. (*a*) Convert $1\frac{1}{2}$ cwt. into kilograms, given that 1 kilogram = 2·2 lb. Give your answer to two decimal places.

(*b*) Calculate to two decimal places how many kilometres are equal to 1·5 miles, given that 1 metre = 39·37 inches and 1 kilometre = 1000 metres.

A6. The capital invested in a business by three partners is £1500, £2000, and £3000 respectively. If the profits during a year amount to £1300, what share did each partner receive?

A7. (*a*) Express the following percentages as fractions:

12 per cent, 54 per cent, 0·5 per cent.

(*b*) Coal is bought at £112. 10*s*. for 50 tons, and is sold at £3 per ton. What is the gain per cent?

A8. During a month a messman took up stores from the Paymaster which exceeded his allowance by £1. 12*s*. 6*d*. and, in addition, made the following outside purchases: 25 tins salmon at 1*s*. 6*d*. per tin; 34 lb. haddock at 1*s*. 7½*d*. per lb.; 48 lb. kippers at 10*d*. per lb.; 4 lb. oranges at 7½*d*. per lb.; 6 lb. apples at 9*d*. per lb.; 60 lb. tomatoes at 1*s*. 2½*d*. per lb.

Find the total costs for the month. If there were 30 men in the mess, find to the nearest penny how much each man would have to pay.

A9. Mess Account for March. 25 men in the Mess.

	£	*s*.	*d*.
Messing allowance	25	10	4
Special dinner vouchers		13	4
Provisions issued on repayment	14	12	3
Mess traps lost		10	8
Canteen bill	11	4	8
Extra expenses during month	2	5	6
Due from messmates for supplies		12	6

Find to the nearest penny the amount each man must pay to keep the mess out of debt.

March, 1942

B1. (*a*) Simplify $\dfrac{4\frac{1}{3}+1\frac{5}{27}}{4\frac{1}{3}-1\frac{5}{27}}$.

(*b*) If $\frac{5}{8}$ of a ship's cargo is worth £10,550, calculate the value of the whole cargo.

B2. (*a*) Simplify $\dfrac{3{\cdot}46\times 2{\cdot}56}{4{\cdot}72}$. Answer to be given to two decimal places only.

(*b*) Express £5. 7*s*. 6*d*. as a decimal fraction of £8. 12*s*.

B3. A ship steams at 25 knots for 10 hours, 28 knots for 6¼ hours, and 20 knots for 3½ hours. How far did she travel, and what was the average speed for the whole time?

B4. What is the area of the walls of a room which is 20 feet long, 16 feet wide, and 10 feet high, and what would be the cost of painting the walls at 8*d*. per sq. yard?

B5. (*a*) Convert $2\frac{1}{2}$ cwt. into kilograms, given that 1 kilogram = 2·2 lb. Give your answer to two decimal places.

(*b*) Calculate to two decimal places how many kilometres are equal to 1 mile, given that 1 metre = 39·37 inches and 1 kilometre = 1000 metres.

B6. The capital invested in a business by three partners is £2500, £3800, and £4600 respectively. If the profits during a year amount to £2180, what share should each partner receive?

B7. (*a*) Express the following percentages as fractions:

28 per cent, 62 per cent, 0·25 per cent.

(*b*) Coal is bought at £135 for 60 tons, and is sold at £2. 14*s*. per ton. What is the gain per cent?

B8. During a month a messman took up stores from the Paymaster which exceeded his allowance by £1. 9*s*. 8*d*. and, in addition, made the following outside purchases: 30 tins salmon at 1*s*. 6*d*. per tin; 24 lb. haddock at 1*s*. $7\frac{1}{2}$*d*. per lb.; 28 lb. kippers at 10*d*. per lb.; 6 lb. oranges at $7\frac{1}{2}$*d*. per lb.; 9 lb. apples at 9*d*. per lb.; 42 lb. tomatoes at 1*s*. $2\frac{1}{2}$*d*. per lb.

Find the total costs for the month. If there were 25 men in the mess, find to the nearest penny how much each man would have to pay.

B9. Mess Account for March. 25 men in the Mess.

	£	*s*.	*d*.
Messing Allowance	24	18	4
Special dinner vouchers		14	6
Provisions issued on repayment	14	18	4
Mess traps lost		11	3
Canteen bill	11	18	2
Extra expenses during month	2	6	8
Due from messmates for supplies		14	0

Find to the nearest penny the amount each man must pay to keep the mess out of debt.

July, 1942

A1. (*a*) Simplify $\frac{3}{7}$ of $\frac{45}{49} \div \frac{2}{3}$ of $\frac{7}{12}$.

(*b*) Of a pole, $\frac{1}{12}$ is painted white, $\frac{7}{30}$ green, $\frac{13}{60}$ red and the remaining 5 feet black. Find the length of the pole.

A2. (*a*) Reduce to decimals $\frac{5}{8}$, $\frac{3}{16}$, $\frac{1}{32}$.

(*b*) Reduce to fractions in lowest terms 0·002, 0·0125, 0·95.

(*c*) Find the value of £15·875.

A3. A tug-of-war team of 8 weigh respectively 10 st. 8 lb., 10 st. 6 lb., 10 st. 12 lb., 11 st. 5 lb., 11 st. 9 lb., 10 st. 7 lb., 11 st. 6 lb. and 14 st. 9 lb. Find the average weight of the team.

A4. In $4\frac{3}{4}$ days a watch loses 1 min. 35 sec. If it is put right at midnight on New Year's Eve, when will it be a quarter of an hour slow?

A5. (*a*) If the rate of exchange is 25·53 francs to the £, what shall I get in francs and centimes (100 centimes equal 1 franc) for £8. 12*s*. 6*d*.?

(*b*) Which pay would you prefer: 5*s*. a day or $6\frac{1}{2}$ francs a day at the above rate of exchange? (Show your working.)

A6. Out of 1500 shells fired, 110 failed to explode. What percentage was properly manufactured?

A7. Prize money of £1280 was divided amongst a boat's crew. A P.O. got 5 shares. A L.S. got 3 shares. An A.B. got 2 shares and an Ord. Seaman got 1 share. There were 2 P.O.s, 2 L.S., 5 A.B.s and 6 Ord. Seamen in the boat. What did each receive?

A8. Twenty-four men were victualled in a mess during June. The following Provisions on Repayment were obtained: 16 lb. beef at $8\frac{1}{2}d$. per lb.; 22 lb. flour at 3*d*. per lb.; 48 lb. bread at $2\frac{1}{4}d$. per lb.; 24 tins milk at 5*d*. per tin; 8 tins salmon at $9\frac{3}{4}d$. per tin; 9 lb. tea at 1*s*. 8*d*. per lb.

The Canteen Bill amounted to £2. 6*s*. $9\frac{1}{2}d$., of which £1. 4*s*. $11\frac{1}{2}d$. was for Private Supplies. Mess traps lost were charged 5*s*. 11*d*.

Make out a Mess account showing the total Mess debt and find what each man pays to keep the mess out of debt.

A9. A Petty Officer's pay was 6*s*. 4*d*. per day. He has three G/C badges for each of which he receives 9*d*. per day. For being a T.G.M. he receives 1*s*. 6*d*. per day. During the quarter ending 30 June he had drawn 10 days' leave allowance of 2*s*. 2*d*. a day and he received £1. 14*s*. 0*d*. for kit upkeep. He allotted £7. 10*s*. 0*d*. per month and drew £11. 10*s*. 0*d*. at the table during the quarter. Find his pay down at the quarter end if 4*s*. 10*d*. was to be deducted for National Health Insurance.

July, 1942

B1. (*a*) Simplify $\frac{3}{8}$ of $\frac{45}{49} \div \frac{3}{7}$ of $\frac{8}{15}$.

(*b*) Of a pole, $\frac{1}{10}$ is painted white, $\frac{7}{30}$ green, $\frac{4}{15}$ red and the remaining 6 feet black. Find the length of the pole.

B2. (*a*) Reduce to decimals $\frac{7}{8}$, $\frac{1}{16}$, $\frac{1}{40}$.

(*b*) Reduce to fractions in lowest terms 0·005, 0·0375, 0·85.

(*c*) Find the value of £13·625.

B3. A tug-of-war team of 8 weigh respectively 10 st. 8 lb., 10 st. 7 lb., 10 st. 11 lb., 11 st. 6 lb., 10 st. 5 lb., 11 st. 9 lb., 10 st. 8 lb. and 13 st. 12 lb. Find the average weight of the team.

B4. In $3\frac{1}{4}$ days a watch loses 1 min. 5 sec. If it is put right at midnight on New Year's Eve, when will it be a quarter of an hour slow?

B5. (*a*) If the rate of exchange is 25·52 francs to the £, what shall I get in francs and centimes (100 centimes equal 1 franc) for £7. 17*s.* 6*d.*?

(*b*) Which pay would you prefer: Half-a-crown a day or $3\frac{1}{2}$ francs a day at the above rate of exchange? (Show your working.)

B6. Out of 1800 shells fired, 120 failed to explode. What percentage was properly manufactured?

B7. Prize money of £1650 was divided amongst a boat's crew. A P.O. got 5 shares. A L.S. got 3 shares. An A.B. got 2 shares and an Ord. Seaman got 1 share. There were 2 P.O.s, 3 L.S., 4 A.B.s and 6 Ord. Seamen in the boat. What did each receive?

B8. Twenty-two men were victualled in a mess during June. The following Provisions on Repayment were obtained: 18 lb. beef at $9\frac{1}{2}$*d.* per lb.; 24 lb. flour at 4*d.* per lb.; 40 lb. bread at $2\frac{1}{4}$*d.* per lb.; 30 tins salmon at $9\frac{3}{4}$*d.* per tin.; 8 lb. tea at 1*s.* 8*d.* per lb.

The Canteen Bill amounted to £2. 7*s.* 4*d.*, of which £1. 7*s.* $6\frac{1}{2}$*d.* was for Private Supplies. Mess traps lost were charged 3*s.* 7*d.*

Make out a Mess account showing the total Mess debt and find what each man pays to keep the mess out of debt.

B9. A Petty Officer's pay was 6*s.* 4*d.* per day. He has two G/C badges for each of which he receives 9*d.* per day. For being a T.G.M. he receives 1*s.* 6*d.* per day. During the quarter ending 30 June he had drawn 12 days' leave allowance of 2*s.* 2*d.* a day and he received £1. 14*s.* 0*d.* for kit upkeep. He allotted £6. 10*s.* 0*d.* per month and drew £12. 15*s.* 0*d.* at the table during the quarter. Find his pay down at the quarter end if 4*s.* 10*d.* was to be deducted for National Health Insurance.

November, 1942

A1. (*a*) Divide the sum of $2\frac{1}{4}$ and $\frac{9}{16}$ by the difference between $1\frac{5}{8}$ and $\frac{7}{12}$.

(*b*) If one pipe would fill a cistern in 3 hours, while another one would fill it in 5 hours, how long would it take to fill the cistern if the two pipes were working together?

A2. (*a*) Simplify $\dfrac{4{\cdot}242-1{\cdot}158}{2{\cdot}65}$. Answer to be given to two decimal places only.

(*b*) Find the value of 0·6 of £10. 5*s.* 5*d.* + 0·75 of £3. 4*s.* 8*d.*

A3. A motorist licensed his car for the period 25 March 1942 to 30 June 1942, and paid £4. 1*s.* 5*d.* road tax, £3. 2*s.* 0*d.* insurance, and £1. 6*s.* 7*d.* for petrol, oil and repairs. For this period he was allowed 8 gallons of petrol only. Find (*a*) cost of running the car per gallon, (*b*) the average cost per mile to the nearest halfpenny, assuming that the car travelled 25 miles on 1 gallon of petrol.

A4. A rectangular plot of ground 60 feet long and 20 feet wide has a path of 2 feet wide running round it. Find the area of the path and the cost of cementing it at 1*s.* 6*d.* per square yard.

A5. At the end of a 30-day month a mess of 25 men had to pay a debt of £2. 11*s.* 8*d.* Of the 25 men, 15 had been in the mess for the whole month, 7 for 20 days, and 3 for 10 days. How much should each man pay?

A6. A cruiser steams on five consecutive days 460, 475, 495, 520 and 510 sea miles. How many sea miles must she steam on the sixth day to make her average daily rate 490 sea miles for the whole period?

A7. (*a*) Express as percentages the fractions $\frac{1}{8}$, $\frac{7}{25}$, 0·625.

(*b*) In an action there were 18 per cent killed, and 33 per cent wounded. If the original strength was 24,500, how many were uninjured?

A8. During a month a messman took up stores from the Paymaster which exceeded his allowance by £1. 6*s.* 7*d.*, and in addition made the following outside purchases: $1\frac{1}{4}$ cwt. new potatoes at $2\frac{1}{4}$*d.* per lb.; 56 lb. new peas at $2\frac{1}{2}$*d.* per lb.; 76 lb. apples at $4\frac{1}{2}$*d.* per lb.; 45 lb. tomatoes at 1*s.* $2\frac{1}{2}$*d.* per lb.; 28 lb. fresh fish at 1*s.* 7*d.* per lb.; 20 lb. strawberries at 1*s.* 1*d.* per lb.

Find the total costs for the month. If there were 25 men in the mess, find to the nearest penny how much each man would have to pay.

A9. A Petty Officer receives 6*s.* 5*d.* per day as pay, 2*s.* 0*d.* per day as Gunner's Mate, and 6*d.* per day for two G/C badges. He has been credited with 10 days' leave allowance at 2*s.* 2*d.* per day, and has been paid £20 during the quarter. His kit upkeep allowance for the quarter amounts to £1. 17*s.* 0*d.* and his quarterly allotment to £21. If his Insurance Contribution is 4*s.* 10*d.*, find the amount due to him for the quarter ending 30 September. (This quarter has 92 days.)

November, 1942

B1. (*a*) Divide the sum of $1\frac{1}{8}$ and $\frac{9}{32}$ by the difference between $3\frac{1}{4}$ and $1\frac{1}{6}$.

(*b*) If one pipe would fill a cistern in 2 hours, while another one would fill it in 3 hours, how long would it take to fill the cistern if the two pipes were working together?

B2. (*a*) Simplify $\frac{3{\cdot}842-2{\cdot}035}{1{\cdot}76}$. Answer to be given to two decimal places only.

(*b*) Find the value of 0·25 of £7. 8*s*. 4*d*. + 0·3 of £5. 0*s*. 10*d*.

B3. A motorist licensed his car for the period 25 March 1942 to 30 June 1942, and paid £3. 8*s*. 9*d*. road tax, £2. 12*s*. 0*d*. insurance, and £1. 5*s*. 3*d*. for petrol, oil and repairs. For this period he was allowed 6 gallons of petrol only. Find (*a*) cost of running the car per gallon, (*b*) the average cost per mile to the nearest halfpenny, assuming that the car travelled 30 miles on 1 gallon of petrol.

B4. A rectangular plot of ground 64 feet long and 25 feet wide has a path 2 feet wide running round it. Find the area of this path and the cost of cementing it at 1*s*. 9*d*. per square yard.

B5. At the end of a 30-day month a mess of 20 men had to pay a debt of £3. 16*s*. 8*d*. Of the 20 men, 10 had been in the mess for the whole month, 6 for 20 days, and 4 for 10 days. How much should each man pay?

B6. A cruiser steams on five consecutive days 450, 482, 510, 480 and 395 sea miles. How many sea miles must she steam on the sixth day to make her average daily rate 480 sea miles for the whole period?

B7. (*a*) Express as percentages the fractions $\frac{1}{3}$, $\frac{7}{20}$, 0·375.

(*b*) In an action there were 15 per cent killed, and 28 per cent wounded. If the original strength was 26,500, how many were uninjured?

B8. During a month a messman took up stores from the Paymaster which exceeded his allowance by £1. 8*s*. 8*d*., and in addition made the following outside purchases: $1\frac{3}{4}$ cwt. new potatoes at $2\frac{1}{4}d$. per lb.; 60 lb. new peas at $2\frac{1}{2}d$. per lb.; 65 lb. apples at $4\frac{1}{2}d$. per lb.; 42 lb. tomatoes at 1*s*. $2\frac{1}{2}d$. per lb.; 25 lb. fresh fish at 1*s*. 7*d*. per lb.; 15 lb. strawberries at 1*s*. 1*d*. per lb.

Find the total costs for the month. If there were 30 men in the Mess, find to the nearest penny how much each man would have to pay.

B9. A Petty Officer receives 6*s*. 10*d*. per day as pay, 2*s*. 0*d*. per day as Gunner's Mate, and 6*d*. per day for two G/C badges. He has been credited with 12 days' leave allowance at 2*s*. 2*d*. per day, and he has been paid £19 during the quarter. His kit upkeep allowance for the quarter amounts to £1. 13*s*. 6*d*. and his quarterly allotment to £21. If his Insurance Contribution is 4*s*. 10*d*., find the amount due to him for the quarter ending 30 September. (This quarter has 92 days.)

	0	1	2	3	4	5	6	7	8	9	1	2	3	4	5	6	7	8	9
10	1000	1005	1010	1015	1020	1025	1030	1034	1039	1044	0	1	1	2	2	3	3	4	4
	3162	3178	3194	3209	3225	3240	3256	3271	3286	3302	2	3	5	6	8	9	11	12	13
11	1049	1054	1058	1063	1068	1072	1077	1082	1086	1091	0	1	1	2	2	3	3	4	4
	3317	3332	3347	3362	3376	3391	3406	3421	3435	3450	1	3	4	6	7	9	10	12	13
12	1095	1100	1105	1109	1114	1118	1122	1127	1131	1136	0	1	1	2	2	3	3	4	4
	3464	3479	3493	3507	3521	3536	3550	3564	3578	3592	1	3	4	6	7	8	10	11	13
13	1140	1145	1149	1153	1158	1162	1166	1170	1175	1179	0	1	1	2	2	3	3	3	4
	3606	3619	3633	3647	3661	3674	3688	3701	3715	3728	1	3	4	5	7	8	10	11	12
14	1183	1187	1192	1196	1200	1204	1208	1212	1217	1221	0	1	1	2	2	3	3	3	4
	3742	3755	3768	3782	3795	3808	3821	3834	3847	3860	1	3	4	5	7	8	9	11	12
15	1225	1229	1233	1237	1241	1245	1249	1253	1257	1261	0	1	1	2	2	3	3	3	4
	3873	3886	3899	3912	3924	3937	3950	3962	3975	3987	1	3	4	5	6	8	9	10	11
16	1265	1269	1273	1277	1281	1285	1288	1292	1296	1300	0	1	1	2	2	3	3	3	4
	4000	4012	4025	4037	4050	4062	4074	4087	4099	4111	1	2	4	5	6	7	9	10	11
17	1304	1308	1311	1315	1319	1323	1327	1330	1334	1338	0	1	1	2	2	2	3	3	3
	4123	4135	4147	4159	4171	4183	4195	4207	4219	4231	1	2	4	5	6	7	8	10	11
18	1342	1345	1349	1353	1356	1360	1364	1367	1371	1375	0	1	1	1	2	2	3	3	3
	4243	4254	4266	4278	4290	4301	4313	4324	4336	4347	1	2	3	5	6	7	8	9	10
19	1378	1382	1386	1389	1393	1396	1400	1404	1407	1411	0	1	1	1	2	2	3	3	3
	4359	4370	4382	4393	4405	4416	4427	4438	4450	4461	1	2	3	5	6	7	8	9	10
20	1414	1418	1421	1425	1428	1432	1435	1439	1442	1446	0	1	1	1	2	2	2	3	3
	4472	4483	4494	4506	4517	4528	4539	4550	4561	4572	1	2	3	4	5	7	8	9	10
21	1449	1453	1456	1459	1463	1466	1470	1473	1476	1480	0	1	1	1	2	2	2	3	3
	4583	4593	4604	4615	4626	4637	4648	4658	4669	4680	1	2	3	4	5	6	8	9	10
22	1483	1487	1490	1493	1497	1500	1503	1507	1510	1513	0	1	1	1	2	2	2	3	3
	4690	4701	4712	4722	4733	4743	4754	4764	4775	4785	1	2	3	4	5	6	7	8	9
23	1517	1520	1523	1526	1530	1533	1536	1539	1543	1546	0	1	1	1	2	2	2	3	3
	4796	4806	4817	4827	4837	4848	4858	4868	4879	4889	1	2	3	4	5	6	7	8	9
24	1549	1552	1556	1559	1562	1565	1568	1572	1575	1578	0	1	1	1	2	2	2	3	3
	4899	4909	4919	4930	4940	4950	4960	4970	4980	4990	1	2	3	4	5	6	7	8	9
25	1581	1584	1587	1591	1594	1597	1600	1603	1606	1609	0	1	1	1	2	2	2	3	3
	5000	5010	5020	5030	5040	5050	5060	5070	5079	5089	1	2	3	4	5	6	7	8	9
26	1612	1616	1619	1622	1625	1628	1631	1634	1637	1640	0	1	1	1	2	2	2	2	3
	5099	5109	5119	5128	5138	5148	5158	5167	5177	5187	1	2	3	4	5	6	7	8	9
27	1643	1646	1649	1652	1655	1658	1661	1664	1667	1670	0	1	1	1	2	2	2	2	3
	5196	5206	5215	5225	5235	5244	5254	5263	5273	5282	1	2	3	4	5	6	7	8	9
28	1673	1676	1679	1682	1685	1688	1691	1694	1697	1700	0	1	1	1	1	2	2	2	3
	5292	5301	5310	5320	5329	5339	5348	5357	5367	5376	1	2	3	4	5	6	7	7	8
29	1703	1706	1709	1712	1715	1718	1720	1723	1726	1729	0	1	1	1	1	2	2	2	3
	5385	5394	5404	5413	5422	5431	5441	5450	5459	5468	1	2	3	4	5	5	6	7	8
30	1732	1735	1738	1741	1744	1746	1749	1752	1755	1758	0	1	1	1	1	2	2	2	3
	5477	5486	5495	5505	5514	5523	5532	5541	5550	5559	1	2	3	4	4	5	6	7	8
31	1761	1764	1766	1769	1772	1775	1778	1780	1783	1786	0	1	1	1	1	2	2	2	3
	5568	5577	5586	5595	5604	5612	5621	5630	5639	5648	1	2	3	3	4	5	6	7	8
32	1789	1792	1794	1797	1800	1803	1806	1808	1811	1814	0	1	1	1	1	2	2	2	2
	5657	5666	5675	5683	5692	5701	5710	5718	5727	5736	1	2	3	3	4	5	6	7	8

The first significant figure and the position of the decimal point must be determined by inspection

	0	1	2	3	4	5	6	7	8	9	1	2	3	4	5	6	7	8
33	1817	1819	1822	1825	1828	1830	1833	1836	1838	1841	0	1	1	1	1	2	2	2
	5745	5753	5762	5771	5779	5788	5797	5805	5814	5822	1	2	3	3	4	5	6	7
34	1844	1847	1849	1852	1855	1857	1860	1863	1865	1868	0	1	1	1	1	2	2	2
	5831	5840	5848	5857	5865	5874	5882	5891	5899	5908	1	2	3	3	4	5	6	7
35	1871	1873	1876	1879	1881	1884	1887	1889	1892	1895	0	1	1	1	1	2	2	2
	5916	5925	5933	5941	5950	5958	5967	5975	5983	5992	1	2	2	3	4	5	6	7
36	1897	1900	1903	1905	1908	1910	1913	1916	1918	1921	0	1	1	1	1	2	2	2
	6000	6008	6017	6025	6033	6042	6050	6058	6066	6075	1	2	2	3	4	5	6	7
37	1924	1926	1929	1931	1934	1936	1939	1942	1944	1947	0	1	1	1	1	2	2	2
	6083	6091	6099	6107	6116	6124	6132	6140	6148	6156	1	2	2	3	4	5	6	7
38	1949	1952	1954	1957	1960	1962	1965	1967	1970	1972	0	1	1	1	1	2	2	2
	6164	6173	6181	6189	6197	6205	6213	6221	6229	6237	1	2	2	3	4	5	6	6
39	1975	1977	1980	1982	1985	1987	1990	1992	1995	1997	0	1	1	1	1	2	2	2
	6245	6253	6261	6269	6277	6285	6293	6301	6309	6317	1	2	2	3	4	5	6	6
40	2000	2002	2005	2007	2010	2012	2015	2017	2020	2022	0	0	1	1	1	1	2	2
	6325	6332	6340	6348	6356	6364	6372	6380	6387	6395	1	2	2	3	4	5	6	6
41	2025	2027	2030	2032	2035	2037	2040	2042	2045	2047	0	0	1	1	1	1	2	2
	6403	6411	6419	6427	6434	6442	6450	6458	6465	6473	1	2	2	3	4	5	5	6
42	2049	2052	2054	2057	2059	2062	2064	2066	2069	2071	0	0	1	1	1	1	2	2
	6481	6488	6496	6504	6512	6519	6527	6535	6542	6550	1	2	2	3	4	5	5	6
43	2074	2076	2078	2081	2083	2086	2088	2090	2093	2095	0	0	1	1	1	1	2	2
	6557	6565	6573	6580	6588	6595	6603	6611	6618	6626	1	2	2	3	4	5	5	6
44	2098	2100	2102	2105	2107	2110	2112	2114	2117	2119	0	0	1	1	1	1	2	2
	6633	6641	6648	6656	6663	6671	6678	6686	6693	6701	1	2	2	3	4	4	5	6
45	2121	2124	2126	2128	2131	2133	2135	2138	2140	2142	0	0	1	1	1	1	2	2
	6708	6716	6723	6731	6738	6745	6753	6760	6768	6775	1	1	2	3	4	4	5	6
46	2145	2147	2149	2152	2154	2156	2159	2161	2163	2166	0	0	1	1	1	1	2	2
	6782	6790	6797	6804	6812	6819	6826	6834	6841	6848	1	1	2	3	4	4	5	6
47	2168	2170	2173	2175	2177	2179	2182	2184	2186	2189	0	0	1	1	1	1	2	2
	6856	6863	6870	6877	6885	6892	6899	6907	6914	6921	1	1	2	3	4	4	5	6
48	2191	2193	2195	2198	2200	2202	2205	2207	2209	2211	0	0	1	1	1	1	2	2
	6928	6935	6943	6950	6957	6964	6971	6979	6986	6993	1	1	2	3	4	4	5	6
49	2214	2216	2218	2220	2223	2225	2227	2229	2232	2234	0	0	1	1	1	1	2	2
	7000	7007	7014	7021	7029	7036	7043	7050	7057	7064	1	1	2	3	4	4	5	6
50	2236	2238	2241	2243	2245	2247	2249	2252	2254	2256	0	0	1	1	1	1	2	2
	7071	7078	7085	7092	7099	7106	7113	7120	7127	7134	1	1	2	3	4	4	5	6
51	2258	2261	2263	2265	2267	2269	2272	2274	2276	2278	0	0	1	1	1	1	2	2
	7141	7148	7155	7162	7169	7176	7183	7190	7197	7204	1	1	2	3	4	4	5	6
52	2280	2283	2285	2287	2289	2291	2293	2296	2298	2300	0	0	1	1	1	1	2	2
	7211	7218	7225	7232	7239	7246	7253	7259	7266	7273	1	1	2	3	3	4	5	6
53	2302	2304	2307	2309	2311	2313	2315	2317	2319	2322	0	0	1	1	1	1	2	2
	7280	7287	7294	7301	7308	7314	7321	7328	7335	7342	1	1	2	3	3	4	5	5
54	2324	2326	2328	2330	2332	2335	2337	2339	2341	2343	0	0	1	1	1	1	1	2
	7348	7355	7362	7369	7376	7382	7389	7396	7403	7409	1	1	2	3	3	4	5	5

The first significant figure and the position of the decimal point must be determined by inspection.

	0	1	2	3	4	5	6	7	8	9	1	2	3	4	5	6	7	8	9
55	2345	2347	2349	2352	2354	2356	2358	2360	2362	2364	0	0	1	1	1	1	1	2	2
	7416	7423	7430	7436	7443	7450	7457	7463	7470	7477	1	1	2	3	3	4	5	5	6
56	2366	2369	2371	2373	2375	2377	2379	2381	2383	2385	0	0	1	1	1	1	1	2	2
	7483	7490	7497	7503	7510	7517	7523	7530	7537	7543	1	1	2	3	3	4	5	5	6
57	2387	2390	2392	2394	2396	2398	2400	2402	2404	2406	0	0	1	1	1	1	1	2	2
	7550	7556	7563	7570	7576	7583	7589	7596	7603	7609	1	1	2	3	3	4	5	5	6
58	2408	2410	2412	2415	2417	2419	2421	2423	2425	2427	0	0	1	1	1	1	1	2	2
	7616	7622	7629	7635	7642	7649	7655	7662	7668	7675	1	1	2	3	3	4	5	5	6
59	2429	2431	2433	2435	2437	2439	2441	2443	2445	2447	0	0	1	1	1	1	1	2	2
	7681	7688	7694	7701	7707	7714	7720	7727	7733	7740	1	1	2	3	3	4	5	5	6
60	2449	2452	2454	2456	2458	2460	2462	2464	2466	2468	0	0	1	1	1	1	1	2	2
	7746	7752	7759	7765	7772	7778	7785	7791	7797	7804	1	1	2	3	3	4	4	5	6
61	2470	2472	2474	2476	2478	2480	2482	2484	2486	2488	0	0	1	1	1	1	1	2	2
	7810	7817	7823	7829	7836	7842	7849	7855	7861	7868	1	1	2	3	3	4	4	5	6
62	2490	2492	2494	2496	2498	2500	2502	2504	2506	2508	0	0	1	1	1	1	1	2	2
	7874	7880	7887	7893	7899	7906	7912	7918	7925	7931	1	1	2	3	3	4	4	5	6
63	2510	2512	2514	2516	2518	2520	2522	2524	2526	2528	0	0	1	1	1	1	1	2	2
	7937	7944	7950	7956	7962	7969	7975	7981	7987	7994	1	1	2	3	3	4	4	5	6
64	2530	2532	2534	2536	2538	2540	2542	2544	2546	2548	0	0	1	1	1	1	1	2	2
	8000	8006	8012	8019	8025	8031	8037	8044	8050	8056	1	1	2	2	3	4	4	5	6
65	2550	2551	2553	2555	2557	2559	2561	2563	2565	2567	0	0	1	1	1	1	1	2	2
	8062	8068	8075	8081	8087	8093	8099	8106	8112	8118	1	1	2	2	3	4	4	5	5
66	2569	2571	2573	2575	2577	2579	2581	2583	2585	2587	0	0	1	1	1	1	1	2	2
	8124	8130	8136	8142	8149	8155	8161	8167	8173	8179	1	1	2	2	3	4	4	5	5
67	2588	2590	2592	2594	2596	2598	2600	2602	2604	2606	0	0	1	1	1	1	1	2	2
	8185	8191	8198	8204	8210	8216	8222	8228	8234	8240	1	1	2	2	3	4	4	5	5
68	2608	2610	2612	2613	2615	2617	2619	2621	2623	2625	0	0	1	1	1	1	1	2	2
	8246	8252	8258	8264	8270	8276	8283	8289	8295	8301	1	1	2	2	3	4	4	5	5
69	2627	2629	2631	2632	2634	2636	2638	2640	2642	2644	0	0	1	1	1	1	1	2	2
	8307	8313	8319	8325	8331	8337	8343	8349	8355	8361	1	1	2	2	3	4	4	5	5
70	2646	2648	2650	2651	2653	2655	2657	2659	2661	2663	0	0	1	1	1	1	1	2	2
	8367	8373	8379	8385	8390	8396	8402	8408	8414	8420	1	1	2	2	3	4	4	5	5
71	2665	2666	2668	2670	2672	2674	2676	2678	2680	2681	0	0	1	1	1	1	1	1	2
	8426	8432	8438	8444	8450	8456	8462	8468	8473	8479	1	1	2	2	3	3	4	5	5
72	2683	2685	2687	2689	2691	2693	2694	2696	2698	2700	0	0	1	1	1	1	1	1	2
	8485	8491	8497	8503	8509	8515	8521	8526	8532	8538	1	1	2	2	3	3	4	5	5
73	2702	2704	2706	2707	2709	2711	2713	2715	2717	2718	0	0	1	1	1	1	1	1	2
	8544	8550	8556	8562	8567	8573	8579	8585	8591	8597	1	1	2	2	3	3	4	5	5
74	2720	2722	2724	2726	2728	2729	2731	2733	2735	2737	0	0	1	1	1	1	1	1	2
	8602	8608	8614	8620	8626	8631	8637	8643	8649	8654	1	1	2	2	3	3	4	5	5
75	2739	2740	2742	2744	2746	2748	2750	2751	2753	2755	0	0	1	1	1	1	1	1	2
	8660	8666	8672	8678	8683	8689	8695	8701	8706	8712	1	1	2	2	3	3	4	5	5
76	2757	2759	2760	2762	2764	2766	2768	2769	2771	2773	0	0	1	1	1	1	1	1	2
	8718	8724	8729	8735	8741	8746	8752	8758	8764	8769	1	1	2	2	3	3	4	5	5
77	2775	2777	2778	2780	2782	2784	2786	2787	2789	2791	0	0	1	1	1	1	1	1	2
	8775	8781	8786	8792	8798	8803	8809	8815	8820	8826	1	1	2	2	3	3	4	4	5

The first significant figure and the position of the decimal point must be determined by inspection.

	0	1	2	3	4	5	6	7	8	9	1	2	3	4	5	6		8
78	2793	2795	2796	2798	2800	2802	2804	2805	2807	2809	0	0	1	1	1	1	1	1
	8832	8837	8843	8849	8854	8860	8866	8871	8877	8883	1	1	2	2	3	3	4	4
79	2811	2812	2814	2816	2818	2820	2821	2823	2825	2827	0	0	1	1	1	1	1	1
	8888	8894	8899	8905	8911	8916	8922	8927	8933	8939	1	1	2	2	3	3	4	4
80	2828	2830	2832	2834	2835	2837	2839	2841	2843	2844	0	0	1	1	1	1	1	1
	8944	8950	8955	8961	8967	8972	8978	8983	8989	8994	1	1	2	2	3	3	4	4
81	2846	2848	2850	2851	2853	2855	2857	2858	2860	2862	0	0	1	1	1	1	1	1
	9000	9006	9011	9017	9022	9028	9033	9039	9044	9050	1	1	2	2	3	3	4	4
82	2864	2865	2867	2869	2871	2872	2874	2876	2877	2879	0	0	1	1	1	1	1	1
	9055	9061	9066	9072	9077	9083	9088	9094	9099	9105	1	1	2	2	3	3	4	4
83	2881	2883	2884	2886	2888	2890	2891	2893	2895	2897	0	0	1	1	1	1	1	1
	9110	9116	9121	9127	9132	9138	9143	9149	9154	9160	1	1	2	2	3	3	4	4
84	2898	2900	2902	2903	2905	2907	2909	2910	2912	2914	0	0	1	1	1	1	1	1
	9165	9171	9176	9182	9187	9192	9198	9203	9209	9214	1	1	2	2	3	3	4	4
85	2915	2917	2919	2921	2922	2924	2926	2927	2929	2931	0	0	1	1	1	1	1	1
	9220	9225	9230	9236	9241	9247	9252	9257	9263	9268	1	1	2	2	3	3	4	4
86	2933	2934	2936	2938	2939	2941	2943	2944	2946	2948	0	0	1	1	1	1	1	1
	9274	9279	9284	9290	9295	9301	9306	9311	9317	9322	1	1	2	2	3	3	4	4
87	2950	2951	2953	2955	2956	2958	2960	2961	2963	2965	0	0	1	1	1	1	1	1
	9327	9333	9338	9343	9349	9354	9359	9365	9370	9375	1	1	2	2	3	3	4	4
88	2966	2968	2970	2972	2973	2975	2977	2978	2980	2982	0	0	1	1	1	1	1	1
	9381	9386	9391	9397	9402	9407	8413	9418	9423	9429	1	1	2	2	3	3	4	4
89	2983	2985	2987	2988	2990	2992	2993	2995	2997	2998	0	0	1	1	1	1	1	1
	9434	9439	9445	9450	9455	9460	9466	9471	9476	9482	1	1	2	2	3	3	4	4
90	3000	3002	3003	3005	3007	3008	3010	3012	3013	3015	0	0	0	1	1	1	1	1
	9487	9492	9497	9503	9508	9513	9518	9524	9529	9534	1	1	2	2	3	3	4	4
91	3017	3018	3020	3022	3023	3025	3027	3028	3030	3032	0	0	0	1	1	1	1	1
	9539	9545	9550	9555	9560	9566	9571	9576	9581	9586	1	1	2	2	3	3	4	4
92	3033	3035	3036	3038	3040	3041	3043	3045	3046	3048	0	0	0	1	1	1	1	1
	9592	9597	9602	9607	9612	9618	9623	9628	9633	9638	1	1	2	2	3	3	4	4
93	3050	3051	3053	3055	3056	3058	3059	3061	3063	3064	0	0	0	1	1	1	1	1
	9644	9649	9654	9659	9664	9670	9675	9680	9685	9690	1	1	2	2	3	3	4	4
94	3066	3068	3069	3071	3072	3074	3076	3077	3079	3081	0	0	0	1	1	1	1	1
	9695	9701	9706	9711	9716	9721	9726	9731	9737	9742	1	1	2	2	3	3	4	4
95	3082	3084	3085	3087	3089	3090	3092	3094	3095	3097	0	0	0	1	1	1	1	1
	9747	9752	9757	9762	9767	9772	9778	9783	9788	9793	1	1	2	2	3	3	4	4
96	3098	3100	3102	3103	3105	3106	3108	3110	3111	3113	0	0	0	1	1	1	1	1
	9798	9803	9808	9813	9818	9823	9829	9834	9839	9844	1	1	2	2	3	3	4	4
97	3114	3116	3118	3119	3121	3122	3124	3126	3127	3129	0	0	0	1	1	1	1	1
	9849	9854	9859	9864	9869	9874	9879	9884	9889	9894	1	1	2	2	3	3	4	4
98	3130	3132	3134	3135	3137	3138	3140	3142	3143	3145	0	0	0	1	1	1	1	1
	9899	9905	9910	9915	9920	9925	9930	9935	9940	9945	0	1	1	2	2	3	3	4
99	3146	3148	3150	3151	3153	3154	3156	3158	3159	3161	0	0	0	1	1	1	1	1
	9950	9955	9960	9965	9970	9975	9980	9985	9990	9995	0	1	1	2	2	3	3	4

The first significant figure and the position of the decimal point must be determined by inspection.

ANSWERS

Exercise I

1. 1365. 2. 1518. 3. 1516. 4. 4635.
5. 14,566. 6. 127,596. 7. 21,189. 8. 81,353.
9. 20,477. 10. 12,152. 11. 32. 12. 42.
13. 307. 14. 105. 15. 2003. 16. 5281.
17. 155. 18. 3418. 19. 21,007. 20. 99.

Exercise II

1. 79,744. 2. 75,744. 3. 116,226. 4. 1,046,576.
5. 28,600,290. 6. 1,861,194. 7. 41,414,282. 8. 68,640,010.
9. 37,489,335. 10. 308,720,124. 11. 52. 12. 24.
13. 437. 14. 981. 15. 274. 16. 3840.
17. 721 (remainder 100). 18. 27 (remainder 7).
19. 79 (remainder 1744). 20. 95 (remainder 911).

Exercise III

1. 10 yd. 2 ft. 3 in. 2. 5 fathoms 2 ft. 7 in.
3. 8 ft. 8 in. 4. 284 in.
5. 16 fathoms 4 ft. 9 in. 6. 87 yd. 1 ft. 6 in.
7. 1500 ft. 8. 325 yd.
9. 12 yd. 10. 7 yd. 1 ft. 2 in.
11. 10 yd. 1 ft. 6 in. 12. 9 fathoms 3 ft. 4 in.
13. 33 shackles 5 fathoms 5 ft. 9 in. 14. 109 fathoms 1 ft. 6 in.
15. 1 ft. 2 in. 16. 2 ft. 9 in.
17. 3 ft. 6 in. 18. 5 ft.
19. 12 fathoms 3 ft. 6 in. 20. 2 ft. 7 in.

Exercise IV

1. 11 hr. 17 min. 24 sec. 2. 1 hr. 32 min. 43 sec.
3. 34 hr. 34 min. 16 sec. 4. 1 hr. 20 min. 41 sec.
5. 9980 sec. 6. 2 hr. 46 min. 40 sec.
7. 46 hr. 8. 8784 hr. 9. 184 days. 10. 91 days.

Exercise V

1. 30 tons 1 cwt. 1 qr. 2. 7 tons 12 cwt. 3 qr.
3. 620 stones. 4. 79 tons 9 cwt.
5. 94 tons 4 cwt. 6. 1 ton 18 cwt. 3 qr.
7. 8 tons 13 cwt. 8. 4 cwt. 3 st. 7 lb.
9. 562 lb. 10. 19 cwt.

Exercise VI

1. 76° 47′ 30″. **2.** 18° 35′ 35″. **3.** 47° 37′ 48″. **4.** 10° 3′ 50″.
5. 12,521 sec. **6.** 5° 28′. **7.** 10° 0′ 1″. **8.** 46° 30′.
9. 1° 59′. **10.** 12° 55′ 35″.

Exercise VII

1. 1349 N.M. **2.** 514 N.M. **3.** 820 N.M.
4. 2nd ship by 556 N.M. **5.** 8014 N.M.
6. 3853 N.M. **7.** 648 N.M. **8.** 4624 N.M.
9. 965 N.M. **10.** 7 S.M.; 480 ft.
11. Both the same, 401,280 ft. **12.** 22 S.M. by 4160 ft.
13. 21 N.M. by 5180 ft. **14.** 2400 yd. **15.** 241 N.M.
16. 25 kt. **17.** 80 tons. **18.** 325 tons.
19. 222 tons. **20.** 6 tons increase. **21.** 15,050 cu. ft.
22. 22 tons 2 cwt. **23.** 214 tons. **24.** 30 tons.

Exercise VIII

1. (*a*) $2\times2\times2\times2$. (*b*) $5\times3\times3$. (*c*) 5×17. (*d*) $2\times2\times2\times3\times7$.
(*e*) 251×1. (*f*) $3\times3\times3\times7\times7\times7$.
2. (*a*) 54. (*b*) 240. (*c*) 126. (*d*) 216. (*e*) 864. (*f*) 792.
3. 1260 fathoms. **4.** 720. **5.** 60 gallons.
6. 15 gallons. **7.** 3 tons 4 cwt. 32 lb.
8. 4320 lb.; yellow 135; red 120; white 160; Dantzig 108 cu. ft.
9. 66,880 yd. **10.** 35 ft.

Exercise IX

1. (*a*) $\frac{3}{4}$. (*b*) $\frac{3}{4}$. (*c*) $\frac{3}{5}$. (*d*) $\frac{1}{3}$. (*e*) $\frac{4}{7}$. (*f*) $\frac{9}{20}$.
(*g*) $\frac{2}{3}$. (*h*) $\frac{8}{9}$. (*i*) $\frac{5}{21}$. (*j*) $\frac{7}{12}$.
2. (*a*) $\frac{7}{4}$. (*b*) $\frac{7}{2}$. (*c*) $\frac{57}{7}$. (*d*) $\frac{34}{5}$. (*e*) $\frac{457}{100}$.
(*f*) $\frac{503}{100}$. (*g*) $\frac{117}{22}$. (*h*) $\frac{197}{15}$. (*i*) $\frac{534}{31}$. (*j*) $\frac{627}{25}$.
3. (*a*) $1\frac{1}{4}$. (*b*) $2\frac{2}{3}$. (*c*) $15\frac{1}{2}$. (*d*) $1\frac{5}{12}$. (*e*) $3\frac{1}{7}$.
(*f*) $5\frac{7}{100}$. (*g*) $26\frac{1}{4}$. (*h*) $42\frac{1}{3}$. (*i*) $9\frac{3}{8}$. (*j*) $7\frac{4}{9}$.
4. (*a*) 20. (*b*) 24. (*c*) 99. (*d*) 51. (*e*) 9, 21.
(*f*) 18, 90. (*g*) 21, 9. (*h*) 26, 91, 9.

Exercise X

1. (*a*) $\frac{7}{10}$. (*b*) $\frac{19}{42}$. (*c*) $\frac{17}{30}$. (*d*) $\frac{53}{60}$. (*e*) $\frac{43}{60}$.
   (*f*) $8\frac{19}{24}$. (*g*) $3\frac{7}{20}$. (*h*) $6\frac{23}{36}$. (*i*) $4\frac{3}{4}$. (*j*) $4\frac{1}{30}$.
2. (*a*) $\frac{4}{21}$. (*b*) $\frac{7}{12}$. (*c*) $\frac{1}{6}$. (*d*) $\frac{7}{36}$. (*e*) $1\frac{1}{2}$.
   (*f*) $1\frac{7}{8}$. (*g*) $1\frac{19}{24}$. (*h*) $3\frac{9}{10}$. (*i*) $\frac{34}{35}$. (*j*) $\frac{13}{14}$.
3. (*a*) $\frac{7}{20}$. (*b*) $1\frac{91}{1000}$. (*c*) $4\frac{1}{3}$. (*d*) $6\frac{5}{8}$. (*e*) $7\frac{29}{30}$.
   (*f*) $2\frac{149}{300}$. (*g*) $1\frac{2}{3}$. (*h*) $\frac{29}{36}$. (*i*) $2\frac{4}{33}$. (*j*) $7\frac{40}{63}$.
4. (*a*) $\frac{11}{42}$. (*b*) $\frac{3}{1000}$. (*c*) $\frac{7}{72}$. (*d*) $\frac{1}{24}$. (*e*) $\frac{1}{84}$. (*f*) $\frac{1}{6}$.

Exercise XI

1. $\frac{5}{6}$; $\frac{4}{5}$; $\frac{3}{4}$; $\frac{7}{10}$; $\frac{2}{3}$.
2. (*a*) 375. (*b*) 250. (*c*) 875. (*d*) $312\frac{1}{2}$.
3. $\frac{4}{15}$.
4. $499\frac{1}{12}$ ft.
5. $\frac{83}{120}$.
6. $\frac{7}{80}$.
7. $\frac{7}{24}$.
8. $\frac{5}{16}$.
9. $\frac{13}{40}$.
10. $\frac{2}{21}$.
11. Starboard; $\frac{3}{35}$.
12. $\frac{5}{12}$.

Exercise XII

1. (*a*) $\frac{6}{11}$. (*b*) $\frac{2}{3}$. (*c*) $\frac{1}{8}$. (*d*) $\frac{4}{15}$. (*e*) $\frac{25}{28}$. (*f*) $\frac{27}{35}$. (*g*) $\frac{1}{16}$. (*h*) $\frac{12}{25}$.
2. (*a*) $\frac{5}{6}$. (*b*) $1\frac{1}{2}$. (*c*) 3. (*d*) $11\frac{1}{4}$. (*e*) $6\frac{2}{3}$. (*f*) 63. (*g*) $2\frac{1}{5}$. (*h*) 3.
3. (*a*) $\frac{1}{4}$. (*b*) 9. (*c*) $19\frac{3}{5}$. (*d*) $5\frac{5}{6}$. (*e*) 36. (*f*) $33\frac{3}{4}$.
4. (*a*) 55. (*b*) 181. (*c*) 234. (*d*) 208.

Exercise XIII

1. (*a*) 7 sec. (*b*) $8\frac{3}{4}$ sec. (*c*) $5\frac{1}{4}$ sec.
2. $52\frac{1}{2}$ tons.
3. 53 in.; $6\frac{1}{4}$ in.; $2\frac{3}{4}$ in.
4. (*a*) 95° F. (*b*) $62\frac{3}{5}$° F. (*c*) $144\frac{1}{2}$° F. (*d*) −13° F. (*e*) 70° C.
   (*f*) 85° C. (*g*) $21\frac{2}{3}$° C. (*h*) −10° C.
5. 20.
6. (*a*) $\frac{4}{15}$. (*b*) $\frac{2}{15}$.
7. $\frac{3}{5}$; 100 tons.
8. 120 fathoms.
9. (*a*) $\frac{1}{3}$. (*b*) 48.
10. (*a*) 24 ft. (*b*) 12 ft.

Exercise XIV

1. (*a*) $\frac{7}{40}$. (*b*) 40. (*c*) $1\frac{1}{4}$. (*d*) $2\frac{1}{3}$. (*e*) $5\frac{1}{3}$. (*f*) 6. (*g*) $5\frac{3}{5}$.
   (*h*) $1\frac{4}{5}$. (*i*) 168. (*j*) $6\frac{2}{3}$.
2. (*a*) $18\frac{1}{2}$. (*b*) $15\frac{1}{2}$. (*c*) $12\frac{1}{2}$. (*d*) $22\frac{1}{2}$.
3. (*a*) $11\frac{1}{2}$. (*b*) $17\frac{3}{4}$. (*c*) $10\frac{1}{3}$. (*d*) $19\frac{2}{3}$.
4. $3\frac{1}{4}$ in.

Exercise XV

1. (*a*) $1\frac{5}{44}$ in. (*b*) $\frac{35}{88}$ in. (*c*) $\frac{63}{88}$ in. (*d*) $\frac{7}{8}$ in.
2. $1\frac{2}{5}$ sec. gaining.
3. $2\frac{1}{5}$ sec. gaining.
4. Midnight 11/12 June.
5. 7 lengths; $\frac{1}{8}$ fathom left.
6. 12 pieces; $1\frac{1}{4}$ in. left.
7. $35\frac{21}{25}$ cu. ft.
8. $35\frac{33}{125}$ cu. ft.
9. $8\frac{21}{25}$ lb. per gal.
10. 9.
11. 1800.
12. $7\frac{1}{2}$ hr.

Exercise XVI

1. (*a*) $\frac{5}{24}$. (*b*) $\frac{3}{40}$. (*c*) $\frac{161}{220}$. (*d*) $\frac{19}{52}$. (*e*) $\frac{25}{102}$. (*f*) $\frac{1}{63,360}$.
2. (*a*) $7\frac{1}{3}$ fathoms. (*b*) 8 tons. (*c*) 94 gal. (*d*) £1. 2*s*. 2*d*. (*e*) $121\frac{1}{2}$ fathoms. (*f*) £29. 8*s*. 0*d*.
3. (*a*) $\frac{1}{15,840}$. (*b*) $\frac{1}{36,480}$. (*c*) $\frac{1}{28,800}$.
4. (*a*) $\frac{1}{3}$ ml. = 1 in. (*b*) $2\frac{1}{2}$ ml. = 1 in. (*c*) $1\frac{2}{5}$ ml. = 1 in.
5. $2\frac{1}{4}$ N.M.
6. $\frac{33}{38}$.
7. (*a*) $\frac{50}{127}$. (*b*) $2\frac{27}{50}$.
8. $14\frac{2}{5}$ knots, $10\frac{4}{5}$ knots.
9. $11\frac{1}{4}$ knots, $14\frac{1}{16}$ knots.
10. $\frac{1}{3}$.
11. $\frac{3}{2000}$.
12. $\frac{84}{145}$.

Exercise XVII

1. $1\frac{31}{60}$.
2. $1\frac{13}{60}$.
3. $1\frac{41}{120}$.
4. $1\frac{4}{15}$.
5. $\frac{2}{3}$.
6. $\frac{3}{8}$.
7. $2\frac{2}{3}$.
8. (*a*) $5\frac{1}{10}$. (*b*) $1\frac{41}{50}$.
9. (*a*) $2\frac{1}{4}$. (*b*) $4\frac{25}{36}$.
10. (*a*) $\frac{11}{18}$. (*b*) $\frac{38}{65}$.
11. (*a*) $\frac{1}{3}$. (*b*) $\frac{66}{115}$.
12. (*a*) $4\frac{1}{8}$. (*b*) 4.

Exercise XVIII

1. 5 ft.
2. $17\frac{1}{2}$ knots.
3. $99\frac{3}{16}$ tons.
4. 1800 tons.
5. 19 sec. slow.
6. 11 sec. fast.
7. 750 tons.
8. 1074 N.M.
9. $\frac{2}{15}$.
10. 10 p.m. 10 April.
11. $\frac{17}{36}$.
12. 176 tons, 220 tons.
13. $20\frac{3}{7}$ min.
14. 16 knots, 20 knots.
15. 8 a.m. 3 April.
16. $12\frac{1}{2}$ knots, $17\frac{1}{2}$ knots.
17. 1280 lb.
18. 8 knots.
19. 1478 lb.
20. $\frac{13}{24}$, 144.

Exercise XIX

1. 2.14 p.m.
2. 5 hr. 50 min.
3. (*a*) 4400 gal. (*b*) 1375 gal.
4. 1 hr. 3 min.
5. 1.30 p.m.
6. 480 gal., 2160 gal.
7. $\frac{13}{24}$, $\frac{1}{6}$.
8. 375 gal. per hr.

Exercise XX

1. (*a*) 21·3328. (*b*) 55·183. (*c*) 85·6191. (*d*) 97·22125. (*e*) 285·233. (*f*) 1·7512. (*g*) 379·9707. (*h*) 405·071. (*i*) 80·767.
2. (*a*) 6·19. (*b*) 4·449. (*c*) 5·208. (*d*) 5·124. (*e*) 1·823. (*f*) 0·2246. (*g*) 88·898. (*h*) 36·423. (*i*) 0·090911.
3. 1·19 in. **4.** 2·59 in.
5. (*a*) 1045·8 mb. (*b*) 931 mb. **6.** (*a*) 40·08 in. (*b*) 2·98 in.

Exercise XXI

1. (*a*) 5·936. (*b*) 0·2782. (*c*) 475·64. (*d*) 0·0117. (*e*) 197·892. (*f*) 0·78694. (*g*) 9·984. (*h*) 2684·834. (*i*) 30·72876. (*j*) 54·74794.
2. (*a*) 6·89 sq. in. (*b*) 13·8125 sq. in. (*c*) 23·4241 sq. in.
3. (*a*) 29·2088 in. (*b*) 28·8956 in. (*c*) 30·189 in.
4. (*a*) 7·08° C. (*b*) 5·1° C. (*c*) −4·8° C. (*d*) −56·28° C.
5. (*a*) 813·2 mb. (*b*) 773·2 mb. (*c*) 613·2 mb. (*d*) 833·2 mb. (*e*) 903·2 mb. (*f*) 948·2 mb.
6. (*a*) 6990 ft. (*b*) 10,896 ft. (*c*) 2604 ft. (*d*) 3396 ft. (*e*) 4530 ft.

Exercise XXII

1. (*a*) 21. (*b*) 31·2. (*c*) 1·2. (*d*) 13. (*e*) 0·00248. (*f*) 2020. (*g*) 81·7. (*h*) 7·8. (*i*) 0·24. (*j*) 0·04.
2. (*a*) 0·081. (*b*) 0·358. (*e*) 0·20. (*d*) 0·747. (*e*) 0·010. (*f*) 0·00. (*g*) 4. (*h*) 62.
3. 6879 N.M. **4.** 14 in. **5.** 21 sec. slow. **6.** 2·47 N.M.

Exercise XXIII

1. (*a*) $\frac{1}{2}$. (*b*) $\frac{19}{20}$. (*c*) $\frac{3}{20}$. (*d*) $\frac{47}{50}$. (*e*) $\frac{7}{8}$. (*f*) $\frac{3}{40}$. (*g*) $1\frac{7}{25}$. (*h*) $5\frac{13}{20}$. (*i*) $2\frac{9}{20}$. (*j*) $1\frac{1}{40}$. (*k*) $2\frac{1}{80}$. (*l*) $10\frac{1}{1000}$. (*m*) $3\frac{63}{125}$. (*n*) $10\frac{3}{8}$. (*o*) $\frac{16}{625}$. (*p*) $3\frac{71}{500}$. (*q*) $6\frac{1}{16}$. (*r*) $2\frac{11}{250}$. (*s*) $1\frac{51}{125}$. (*t*) $\frac{11}{10,000}$.
2. (*a*) 0·125. (*b*) 0·625. (*c*) 0·1875. (*d*) 0·3125. (*e*) 0·6875. (*f*) 0·28125. (*g*) 0·71875. (*h*) 0·857. (*i*) 0·462. (*j*) 0·667. (*k*) 0·556. (*l*) 0·412. (*m*) 0·733. (*n*) 0·083. (*o*) 3·143.

Exercise XXIV

1. (*a*) 0·893. (*b*) 1·12. **2.** (*a*) 0·919. (*b*) 1·088.
3. (*a*) 1·175. (*b*) 0·851. **4.** £2. 15*s*. 2*d*.
5. £4·633. **6.** 0·208. **7.** 0·433. **8.** 0·914.
9. 0·868. **10.** 0·556. **11.** 0·03. **12.** 8·439.

Exercise XXV

1. 21·6. **2.** 0·185. **3.** 0·138. **4.** 1·033.
5. 4·25. **6.** 0·916. **7.** 6·12. **8.** 0·025.
9. 0·814. **10.** 4·166.

Exercise XXVI

1. (*a*) 16,758 tons. (*b*) 3,724,000 cu. ft.
2. (*a*) 7640 tons. (*b*) 764,000 cu. ft.
3. (*a*) 3950 tons. (*b*) 395,000 cu. ft. **4.** 4495 tons.
5. (*a*) 3357 tons. (*b*) 7460 tons. **6.** 1827 tons.
7. (*a*) 16,250 tons. (*b*) 10,111 tons. **8.** 7012 tons.
9. 205,200 cu. ft. **10.** 88,000 cu. ft.
11. (*a*) 0·28. (*b*) 2240 tons. (*c*) 1764 tons. (*d*) 0·225.
12. 11,000 tons. **13.** 8400 tons.
14. (*a*) 0·977. (*b*) 0·984.
15. (*a*) 357,000 cu. ft. (*b*) 365,568 cu. ft. (*c*) 0·976.
16. 175,000 cu. ft., decrease. **17.** (*a*) 11,200 tons. (*b*) 9408 cu. ft.
18. 35·56 cu. ft. **19.** (*a*) 5000 tons. (*b*) 175,000 cu. ft. **20.** 4185 tons.

Exercise XXVII

1. (*a*) $\frac{1}{1000}$, 0·001. (*b*) $\frac{1}{1000}$, 0·001. (*c*) $\frac{1}{100}$, 0·01. (*d*) $\frac{3}{10}$, 0·3.
(*e*) $\frac{1}{250}$, 0·004. (*f*) $\frac{1}{2500}$, 0·0004.
2. (*a*) 5·004. (*b*) 1·156. (*c*) 0·056. (*d*) 0·75. (*e*) 0·0024. (*f*) 0·00446.
3. (*a*) 246,200. (*b*) 2·613. (*c*) 4·28. (*d*) 22,000. (*e*) 0·76.
(*f*) 0·0466. (*g*) 3,002,000. (*h*) 0·06675. (*i*) 0·000007. (*j*) 2400.
4. (*a*) 8·27 m. (*b*) 420·724 m. (*c*) 197,561 mm. (*d*) 0·9822 Km.
(*e*) 2721 gm. (*f*) 4,016,500 c.c.

Exercise XXVIII

1. (*a*) 101·4 Km. (*b*) 344·4 Km. (*c*) 130·7 Km. (*d*) 160·9 Km.
2. (*a*) 62·1 ml. (*b*) 353·1 ml. (*c*) 20·1 ml. (*d*) 111·1 ml.
3. *A* by 0·92 ft. **4.** 1·851 Km.
5. (*a*) 220. (*b*) 132. (*c*) 99. (*d*) 550.
6. (*a*) 4546. (*b*) 2727·6. (*c*) 2045·7. (*d*) 19·7.
7. 166$\frac{2}{3}$ min. **8.** 273 gal.
9. Pump *A* by 0·28 gal. per min. **10.** 78·9 kgm. **11.** 718·7 kgm.

12. Rope circumference 1·905 cm.; breaking load 1·727 tonnes.
13. Tay by 1987·68 ft.
14. 5348 fathoms.
15. 9562·8 m.
16. (*a*) 8884·4 m. (*b*) 491 fathoms.
17. 14,300 ft.
18. 13,102 tons; 158 ft.; 9 ft.; 3 ft.
19. 15 ft. 4 in.
20. 1 ft. 8 in. by the stern.
21. 24 ft. 10 in.
22. 6006 tonnes.
23. 10,040 tons.
24. 10,631 tons.
25. $1\frac{9}{10}$ sec.

Exercise XXIX

1. (*a*) 722 tons. (*b*) 37,553 cu. ft.
2. (*a*) 29,297 cu. ft. (*b*) 18,644 cu. ft.
3. 12,571 cu. ft.; 1600 bags.
4. 9714 cu. ft.; 194 tons.
5. 88 tons; 4776 cu. ft.
6. 312 cu. ft.
7. 23 knots.
8. 2*s*. 0*d*.
9. 11*s*. 4*d*.
10. 3*s*. 0*d*.

Exercise XXX

1. 21.87 dollars.
2. £2. 3*s*. $9\frac{1}{2}$*d*.
3. 46 rupees 10 annas.
4. £1. 16*s*. 9*d*.
5. (*a*) 2*s*. 4*d*. (*b*) £4. 7*s*. 10*d*.
6. £2. 3*s*. $6\frac{1}{2}$*d*.
7. £2. 4*s*. $3\frac{1}{2}$*d*.
8. 819 piastres.
9. £5. 12*s*. 0*d*.
10. (*a*) £1. 19*s*. 11*d*. (*b*) 79·50 pesetas.
11. 49·14 roubles.
12. £1. 14*s*. 11*d*.
13. 130 piastres.
14. Nil.
15. 17.60 American dollars.

Exercise XXXI

1. 105 yd.
2. 13·6125 tons.
3. 10·6 tons.
4. 223 tons.
5. 945 feet.
6. $\frac{1}{8}$ N.M.
7. 34 tons; 5 fathoms.
8. 0·9 in.

Exercise XXXII

1. 2 fathoms.
2. 3 fathoms.
3. 3 ft.
4. 4000 fathoms.
5. 4350 fathoms.
6. 3·75 N.M.
7. 187 fathoms.
8. 1·674 N.M.
9. 1·6 sec.
10. 429 fathoms per N.M.

Exercise XXXIII

1. 23 ft. $7\frac{1}{2}$ in.
2. 10 fathoms.
3. 45 sec.
4. 12·1 knots.

Exercise XXXIV

1. 27 sec.
2. 4 hr.; 92 N.M.; 110 N.M.
3. 3 hr. 36 min.
4. 2 min. 15 sec.
5. $32\frac{2}{3}$ sec.
6. 11 45 hr.
7. 45 min.
8. 18 00 hr.; 120 N.M.

9. 16 knots. 10. 40 min.; 19 knots. 11. 25 min.
12. 1000 yd. per min. closing; 12 08 hr. 13. 1 N.M.
14. 10 N.M. 15. 1 N.M. 16. 2 cables.
17. 17 15 hr. 18. 33¾ N.M. 19. 9 N.M.; 8 N.M.
20. 13 12 hr.; 48 N.M.

Exercise XXXV

1. 1005·9. 2. 16½ knots. 3. 29·72. 4. 46° 23′ N.
5. 3° 30′ N. 6. 11 hr. 32 min. 05 sec.; 54° 20′ 54″.
7. 15·1 knots. 8. 186 gross. 9. 72 lb. 10. 11 st. 5 lb.
11. 8 st. 9 lb. 12. 12·25 ft.

Exercise XXXVI

1. (*a*) 16½. (*b*) 40. (*c*) 32. (*d*) 29·05. 2. 1:54,720.
3. 3·15 in. 4. (*a*) 1·02. (*b*) 0·98.

Exercise XXXVII

1. (*a*) 0·49. (*b*) 0·74. (*c*) 1·98. 2. (*a*) 2·4 cwt. (*b*) 6·15 cwt.
3. 18 cwt.; 2 tons 8 cwt. 4. 2 tons 16 cwt.

Exercise XXXVIII

1. 1½ cwt. 2. 15 cwt.
3. Fourfold purchase. 4. No, 84 lb.

Exercise XXXIX

1. 30 ft.; 50 ft.; 70 ft.
2. 33 cwt. copper; 3¾ cwt. tin; ¾ cwt. zinc.
3. 10 in.; 2 ft. 1 in.; 3 ft. 9 in. 4. 8:19.
5. 11*s*. 3*d*.; 15*s*. 9*d*.; £1. 7*s*. 0*d*. 6. 20 ft.
7. 1 ft. 8. 9 ft. 4 in.; 2 ft. 4 in.
9. 540 tons, 315 tons, 225 tons. 10. £55, £35, £25, £20.

Exercise XL

1. (*a*) 16. (*b*) 17·1. (*c*) 19·3. (*d*) 14·8.
2. (*a*) 28·5 N.M. (*b*) 41·9 N.M. (*c*) 97·4 N.M. (*d*) 324·5 N.M.
3. 1260 N.M.
4. (*a*) 11 hr. 15 min. (*b*) 17 hr. 17 min. (*c*) 13 hr. 11 min. (*d*) 15 hr. 1 min.
5. 128 tons; 3 days 18 hr. 6. (*a*) 27 knots. (*b*) 27 hr.
7. 434 francs. 8. (*a*) 11⅓ rupees. (*b*) £3. 3*s*. 0*d*.
9. 504 lb. 10. (*a*) $2\frac{3}{7}$. (*b*) $1\frac{5}{7}$. (*c*) $2\frac{1}{7}$.

Exercise XLI

1. 15 days. 2. 2 days. 3. $6\frac{3}{5}$ knots. 4. 3 knots.
5. 240. 6. $9\frac{1}{3}$ oz. 7. $18\frac{3}{4}$ knots. 8. 144 tons.
9. $\frac{1}{4}$ lb. 10. 0·475 in. 11. $\frac{3}{4}$ pint. 12. 16 days.
13. 1600. 14. 6 men.

Exercise XLII

1. (*a*) 50%. (*b*) 25%. (*c*) 75%. (*d*) $12\frac{1}{2}$%. (*e*) $87\frac{1}{2}$%.
2. (*a*) 18%. (*b*) 210%. (*c*) 34%. (*d*) 70%. (*e*) 6·5%. (*f*) 207%.
3. (*a*) 33·3%. (*b*) 83·3%. (*c*) 22·2%. (*d*) 23·0%. (*e*) 7·8%. (*f*) 65·9%.
4. (*a*) $\frac{3}{20}$. (*b*) $\frac{18}{25}$. (*c*) $\frac{1}{16}$. (*d*) $1\frac{1}{40}$. (*e*) $\frac{3}{80}$. (*f*) 1.
5. (*a*) 0·16. (*b*) 0·025. (*c*) 1·25. (*d*) 0·0375. (*e*) 0·142. (*f*) 0·0105.
6. 4·04%. 7. 57·2 lb. 8. 290 fathoms.
9. 0·18%. 10. (*a*) 2·86%. (*b*) 2·78%.
11. 85% copper, $12\frac{1}{2}$% zinc, $2\frac{1}{2}$% tin.
12. 21 cwt. copper, 4·2 cwt. zinc, 2·8 cwt. tin.

Exercise XLIII

1. 14·3%. 2. 9·1%. 3. 30 knots. 4. 10 ft.
5. 105 lb. wt. 6. 300 lb. 7. 22 lb. wt. 8. 90 lb.

Exercise XLIV

1. (*a*) £111. 6*s*. 6*d*. (*b*) £27. 7*s*. 6*d*.
2. (*a*) 6*s*. 9*d*. per day. (*b*) £47. 18*s*. 6*d*.
3. £29. 9*s*. 4*d*. 4. £39. 13*s*. 7*d*. 5. 67 dollars.
6. £9. 18*s*. 11*d*. 7. £9. 12*s*. 10*d*. 8. £9. 5*s*. 11*d*.

Exercise XLV

1. (*a*) 8*s*. 0*d*. (*b*) 6*d*.; balance 1*s*. 0*d*.
2. (*a*) £1. 15*s*. 6*d*. (*b*) 4*s*. 6*d*. 3. (*a*) 3*s*. 11*d*. (*b*) 15*s*. 1*d*.
4. (*a*) 12*s*. 6*d*. (*b*) 8*d*.
5. (*a*) 7*s*. 0*d*. (*b*) 1*s*. 10*d*.; balance 10*s*. 10*d*.
6. £6. 15*s*. $1\frac{1}{2}$*d*. 7. £1. 15*s*. 11*d*. 8. 4*s*. 2*d*.
9. £6. 14*s*. 3*d*. 10. £1. 4*s*. $3\frac{1}{4}$*d*.; 5*s*. 6*d*.

Exercise XLVI

1. (*a*) 2. (*b*) 4. (*c*) 6. (*d*) 8. (*e*) 9.

2. (*a*) 29. (*b*) 39. (*c*) 97. (*d*) 38. (*e*) 93.

3. (*a*) 5·7. (*b*) 1·802. (*c*) 13·29. (*d*) 2·708. (*e*) 0·1273.
(*f*) 1·012. (*g*) 0·01779. (*h*) 3·001.

4. (*a*) 13·21. (*b*) 1·707. (*c*) 0·1987. (*d*) 1·734. (*e*) 3·326.
(*f*) 0·1741. (*g*) 54·54. (*h*) 105·0.

Exercise XLVII

1. (*a*) 3·64 N.M. (*b*) 7·71 N.M. (*c*) 10·91 N.M. (*d*) 12·06 N.M.

2. (*a*) 170 ft. (*b*) 333 ft. (*c*) 681 ft.

3. (*a*) 12·6 N.M. (*b*) 13·2 N.M. **4.** 10 ft.

5. 18 N.M. **6.** 25 N.M. **7.** 169 ft.

8. 12 ft. **9.** 12·4 N.M. **10.** 23·4 N.M.

Exercise XLVIII

1. 85 sq. ft. **2.** $7\frac{1}{2}$ sq. yd. **3.** 10 gal. **4.** 13 lb.

5. 87 lb. **6.** $62\frac{1}{2}$ sq. yd. **7.** 18. **8.** 45 sq. yd.

9. (*a*) 2106 sq. ft. (*b*) $31\frac{1}{2}$ fathoms. (*c*) 315 lb. **10.** 58 ft.

Exercise XLIX

1. 38. **2.** 41. **3.** 15. **4.** 68. **5.** 57.

6. 89. **7.** 127. **8.** 114. **9.** 293. **10.** 167.

Exercise L

1. 64 sq. ft. **2.** 3 ft. **3.** 6000 sq. ft.

4. (*a*) 1920 cu. ft., (*b*) 12,000 gal., (*c*) $54\frac{6}{7}$ tons.

5. (*a*) 110 tons. (*b*) 100 tons. **6.** 50 ft.

7. 7120 lb. **8.** 6408 lb. **9.** 45 lb.

10. (*a*) 46·7 cu. ft./ton; +16. (*b*) 61·0 lb./cu. ft.; +3.
(*c*) 38·0 lb./cu. ft.; +26. (*d*) 72·3 cu. ft./ton; +33.
(*e*) 49·8 cu. ft./ton; +19. (*f*) 71·0 lb./cu. ft.; −7.
(*g*) 53 lb./cu. ft.; 42·3 cu. ft./ton. (*h*) 84 lb./cu. ft.; 26·7 cu. ft./ton.
(*j*) 60·0 lb./cu. ft.; +4. (*k*) 74·7 cu. ft./ton; +34.

Exercise LI

1. 140 tons. **2.** 68 tons. **3.** 130 tons. **4.** 200 tons.

5. 483 tons. **6.** 150 ft. **7.** 8 ft. **8.** 24 ft.

9. 155 ft. **10.** 9·1 ft.

Exercise LII

	Length	Breadth	Depth	Capacity
1.	9 ft.	4 ft. 0 in.	1 ft. 6 in.	34 cu. ft.
2.	26 ft.	8 ft. 0 in.	3 ft. 4 in.	419 cu. ft.
3.	23 ft.	7 ft. 6 in.	3 ft. 0 in.	315 cu. ft.
4.	18 ft.	6 ft. 3 in.	2 ft. 6 in.	170 cu. ft.
5.	30 ft.	9 ft. 0 in.	3 ft. 9 in.	612 cu. ft.
6.	16 ft.	5 ft. 9 in.	2 ft. 3 in.	128 cu. ft.
7.	27 ft.	8 ft. 3 in.	3 ft. 5 in.	463 cu. ft.
8.	40 ft.	11 ft. 6 in.	4 ft. 9 in.	1333 cu. ft.
9.	15 ft.	5 ft. 6 in.	2 ft. 2 in.	109 cu. ft.
10.	32 ft.	9 ft. 6 in.	3 ft. 11 in.	728 cu. ft.

Exercise LIV

1. (*a*) $40\frac{5}{8}$ gal. (*b*) $9\frac{3}{8}$ cu. ft.
2. (*a*) 672 lb. per sq. ft. (*b*) 11 ft. 3 in.
3. (*a*) £16. (*b*) 50 days.
4. (*a*) 17 ft. 1 in. (*b*) 3840 tons.

Exercise LV

1. (*a*) 11·2 ft. (*b*) 09 54 hr.
2. (*a*) 857 I.H.P. (*b*) 10·5 knots.
3. (*a*) 19·3 tons. (*b*) 13 knots.
4. (*a*) 1190 tons. (*b*) 11·3 ft.

EDUCATIONAL TESTS

July, 1940

A1. (*a*) $\frac{1}{72}$. (*b*) 1 hr.

A2. (*a*) 0·042. (*b*) 0·1734.

A3. 10 hr. 40 min.

A4. £40; £28; £20; £12.

A5. £49. 17*s*. 9*d*.

A6. Havre to Paris by 3140 yd.

A7. $26\frac{6}{19}$ %.

A8. £9. 5*s*. $6\frac{1}{2}$*d*.

A9. Deficit £3. 8*s*. $11\frac{1}{2}$*d*.

B1. (*a*) $\frac{7}{68}$. (*b*) £12. 12*s*. 0*d*.

B2. (*a*) 0·039. (*b*) 4·342.

B3. 8 hr.

B4. £32; £24; £16; £8.

B5. £57. 5*s*. 3*d*.

B6. Exeter to London by 1680 yd.

B7. $33\frac{1}{3}$ %.

B8. £9. 4*s*. $11\frac{1}{2}$*d*.

B9. Credit £1. 8*s*. $3\frac{1}{2}$*d*.

NOVEMBER, 1940

A1. (*a*) $4\frac{9}{19}$. (*b*) £2000.
A2. (*a*) 1·017. (*b*) £1. 0*s*. $3\frac{1}{4}$*d*.
A3. 72 lb.
A4. 28 knots, 23 hr. 20 min.
A5. (*a*) $33\frac{1}{3}$ gal. (*b*) 1800 men.
A6. £4000; £2400; £1600.
A7. (*a*) $12\frac{1}{2}$ %; 75 %; 37·5 %. (*b*) $33\frac{1}{3}$ %.
A8. £10. 16*s*. 4*d*.
A9. £4. 19*s*. 6*d*.

B1. (*a*) 4. (*b*) £2100.
B2. (*a*) 1·256. (*b*) £1. 7*s*. 2*d*.
B3. 72 lb.
B4. 27 knots, $30\frac{6}{7}$ hr.
B5. (*a*) 20 gal. (*b*) 750 men.
B6. £2400; £2000; £1200.
B7. (*a*) 25 %; 85 %; 62·5 %. (*b*) 20 %.
B8. £9. 19*s*. 9*d*.
B9. £5. 10*s*. 6*d*.

MARCH, 1941

A1. (*a*) $1\frac{3}{5}$. (*b*) $\frac{2}{3}$.
A2. (*a*) 0·000001. (*b*) 2·56.
A3. 150.
A4. (*a*) $\frac{3}{16}$; 2 %; 0·017. (*b*) 55·97 % Ceylon, 44·03 % China.
A5. $10\frac{5}{7}$ ft.
A6. 12 st. 10 lb.
A7. £3. 6*s*. 0*d*.
A8. £7. 11*s*. 4*d*.
A9. £5. 19*s*. 4*d*.

B1. (*a*) $1\frac{1}{9}$. (*b*) $\frac{13}{24}$.
B2. (*a*) 20. (*b*) 4·25.
B3. 54.
B4. (*a*) $\frac{2}{25}$; 0·024; 1 %. (*b*) 56·59 % Ceylon, 43·41 % China.
B5. $10\frac{10}{11}$ acres.
B6. 13 st. 2 lb.
B7. £2. 10*s*. 8*d*.
B8. £8. 13*s*. 4*d*.
B9. £7. 1*s*. 2*d*.

JULY, 1941

A1. (*a*) $1\frac{1}{2}$. (*b*) 27 ft.
A2. (*a*) 1·329. (*b*) 0·425, 0·44, $\frac{11}{25}$.
A3. 36 m.p.h., 8 miles.
A4. £264, £198, £110, £44.
A5. £2. 12*s*. 1*d*.
A6. 2·54 cm., 619·35 sq. cm.
A7. (*a*) $33\frac{1}{3}$ %, 72 %, 32·5 %. (*b*) £45.
A8. 9*s*. $4\frac{3}{4}$*d*. (or 9*s*. 5*d*. to nearest penny).
A9. $9\frac{1}{2}$*d*. (leaving $2\frac{1}{2}$*d*. in mess funds).

B1. (*a*) $2\frac{2}{3}$. (*b*) 50 ft.
B2. (*a*) 1·142. (*b*) 0·4125, 0·48, $\frac{12}{25}$.
B3. 48 m.p.h., $2\frac{1}{4}$ miles.
B4. £220, £176, £110, £44.
B5. £2. 18*s*. 4*d*.
B6. 2·54 cm., 903·22 sq. cm.
B7. (*a*) $12\frac{1}{2}$ %, $83\frac{1}{3}$ %, 66·5 %. (*b*) £96.
B8. £9. 18*s*. 7*d*.; 8*s*. $3\frac{1}{2}$*d*. (leaving 5*d*. in mess funds).
B9. 2*s*. 1*d*. (leaving 3*d*. in mess funds).

November, 1941

A1. (*a*) $\frac{4}{25}$. (*b*) $\frac{1}{16}$.
A2. (*a*) 70·03. (*b*) 0·3125.
A3. 300.
A4. (*a*) 5*s*. 0*d*. (*b*) £1. 1*s*. 1*d*.
A5. (*a*) $\frac{1}{50}, \frac{1}{20,000}, \frac{1}{10}, \frac{1}{8}$. (*b*) 0·0625, 0·0006, 0·01, 0·3375.
A6. £2000, £80.
A7. $146\frac{1}{4}$ yd. per min.; 138·75 metres per min.; Egyptian by 5·49 yd.
A8. £5. 12*s*. 2*d*.
A9. 11*s*. $11\frac{1}{2}$*d*.

B1. (*a*) $\frac{3}{14}$. (*b*) $\frac{5}{56}$.
B2. (*a*) 260·09. (*b*) 0·4375.
B3. 200.
B4. (*a*) 7*s*. 0*d*. (*b*) 13*s*. 5*d*.
B5. (*a*) $\frac{1}{25}, \frac{1}{12,500}, \frac{7}{40}, \frac{4}{5}$. (*b*) 0·03125, 0·0014, 0·02, 0·5525.
B6. £3000, £180.
B7. 150 yd. per min.; 145·75 metres per min.; Egyptian by 9·39 yd.
B8. £7. 11*s*. 2*d*.
B9. 12*s*. $0\frac{1}{2}$*d*. (leaving 4*d*. in mess funds).

March, 1942

A1. (*a*) $4\frac{5}{22}$. (*b*) £13,100.
A2. (*a*) 2·26. (*b*) 0·375.
A3. 324 N.M., $22\frac{14}{15}$ knots.
A4. 88 sq. yd., £3. 6*s*. 0*d*.
A5. (*a*) 76·36 kgm. (*b*) 2·41 Km.
A6. £300, £400, £600.
A7. (*a*) $\frac{3}{25}, \frac{27}{50}, \frac{1}{200}$. (*b*) $33\frac{1}{3}$%.
A8. £12. 4*s*. 9*d*.; 8*s*. 2*d*. (leaving 3*d*. in mess funds).
A9. 1*s*. 6*d*.

B1. (*a*) $1\frac{64}{85}$. (*b*) £16,880.
B2. (*a*) 1·87. (*b*) 0·625.
B3. 495 N.M., $25\frac{7}{79}$ knots.
B4. 80 sq. yd., £2. 13*s*. 4*d*.
B5. (*a*) 127·27 kgm. (*b*) 1·60 Km.
B6. £500, £760, £920.
B7. (*a*) $\frac{7}{25}, \frac{31}{50}, \frac{1}{400}$. (*b*) 20%.
B8. £9. 18*s*. 3*d*.; 8*s*. 0*d*. (leaving 1*s*. 9*d*. in mess funds).
B9. 2*s*. 9*d*.

July, 1942.

A1. (*a*) $1\frac{20}{2401}$. (*b*) $10\frac{5}{7}$ ft.
A2. (*a*) 0·625, 0·1875, 0·03125. (*b*) $\frac{1}{500}, \frac{1}{80}, \frac{19}{20}$. (*c*) £15. 17*s*. 6*d*.
A3. 11 st. 6 lb.
A4. Midnight Feb. 14/15.
A5. (*a*) 220 fr. 19 centimes. (*b*) $6\frac{1}{2}$ fr.
A6. $92\frac{2}{3}$%.
A7. £200, £120, £80, £40.
A8. £4. 5*s*. 1*d*, 3*s*. 7*d*. (leaving 11*d*. in mess funds).
A9. £14. 8*s*. 5*d*.

B1. (*a*) $1\frac{227}{418}$. (*b*) 15 ft.
B2. (*a*) 0·875, 0·0625, 0·025. (*b*) $\frac{1}{200}, \frac{3}{80}, \frac{17}{20}$. (*c*) £13. 12*s*. 6*d*.
B3. 11 st. 3 lb.
B4. Midnight Feb. 14/15.
B5. (*a*) 200 fr. 97 centimes. (*b*) $3\frac{1}{2}$ fr.
B6. $93\frac{1}{3}$%.
B7. £250, £150, £100, £50.
B8. £4. 10*s*. 10*d*., 4*s*. 2*d*. (leaving 10*d*. in mess funds).
B9. £12. 19*s*. 6*d*.

November, 1942

A1. (*a*) $2\frac{7}{10}$. (*b*) $1\frac{7}{8}$ hr. **A2.** (*a*) 1·16. (*b*) £8. 11*s*. 9*d*.

A3. (*a*) £1. 1*s*. 3*d*. (*b*) 10*d*. **A4.** 336 sq. ft.; £2. 16*s*. 0*d*.

A5. 15 pay 2*s*. 6*d*.; 7 pay 1*s*. 8*d*.; 3 pay 10*d*. **A6.** 480.

A7. (*a*) $12\frac{1}{2}$ %, 28 %, 62·5 %. (*b*) 12,005.

A8. £10. 13*s*. $4\frac{1}{2}$*d*., 8*s*. 7*d*. **A9.** £2. 14*s*. 2*d*.

B1. (*a*) $\frac{27}{40}$. (*b*) $1\frac{1}{5}$ hr. **B2.** (*a*) 1·02. (*b*) £3. 7*s*. 4*d*.

B3. (*a*) £1. 4*s*. 4*d*. (*b*) $9\frac{1}{2}$*d*. **B4.** 372 sq. ft.; £3. 12*s*. 4*d*.

B5. 10 pay 5*s*. 0*d*.; 6 pay 3*s*. 4*d*.; 4 pay 1*s*. 8*d*. **B6.** 563.

B7. (*a*) $33\frac{1}{3}$ %, 35 %, 37·5 %. (*b*) 15,105.

B8. £10. 8*s*. $10\frac{1}{2}$*d*., 7*s*. 0*d*. **B9.** £5. 13*s*. 4*d*.

INDEX OF NAUTICAL TERMS

Printed in the United States
By Bookmasters